STRESS

"the logistic curve"

'splode

I scream

I sigh

LOGISTICS

SPEED LIMIT

$$\sum_{k=1}^{\infty} \frac{45}{2^k}$$

W9-BBU-601

popularity

HOW DO YOUR RIE

a derivative runs through it

$n=1$

U.S.S. $\frac{d^2}{dt^2}$

ROLLER COS-TER

0D

1D

2D

3D

4D

$\frac{d\,Moon}{dt} > 0$

Q: Where

A: At

$\frac{d\,Moon}{dt} < 0$

Farce

$$\sum_{i=0}^{\infty} \frac{x^i}{i!} =$$

(Taylor Series)

1. Self-titled
2. Fearless
3. Speak Now
4. Red
5. 1989
6. Reputation

THE LIMIT DO NOT EXIST

WENT TO DONTIST

$$\lim_{h \to 0} \frac{f(x+h) - f(x)}{h}$$

THE WEIERSTRASS

vectors

WENT TO ORTHO- DONTIST

THE CONSTANT OF INTEGRATION IS LIKE ANTI-RIDING A BICYCLE:

(nowhere-differentiable dance moves)

arrow moving 100mph

SPOT THE DIFFERENCE

arrow not moving

YOU ALWAYS FORGET IT

$$\int \cos(x)\,dx = \sin x + \frac{death}{and} \; taxes$$

REVENUE

HA HA HA

TEGRAL
ALCULUS
FIC NAMES
ULOUS

"Even a passing moment has its fertile past."

0%

100%

MTR

SQUEE

THE FUN OUT

CHANGE
IS THE ONLY
CONSTANT

$$\nabla T_{+} f$$

$$\sum_{n=1} \frac{\delta}{2} H_i^{M} c_s \frac{P}{a} c_0 D_z \qquad c_2 = c_{s+2} = \frac{}{(s+1)(s+2)} c_s = \frac{}{6 \cdot 7} \cdot 0 = 0$$

$$\frac{Q(p-1)}{2p} H^M f_0 N$$

$$(x) \log p(x) \qquad F_0 N = \sum_{c=1} D_{i,\nu,d,(1+d)}$$

$$\frac{V}{\times T} - \nu \; V - D$$

$$\frac{D}{\times 2} S_i + c_2^\nu D_n + \frac{q H_i^\nu}{2} (m(1 - \frac{\nu}{p}) - 1 - 2 \frac{D}{p})]$$

$$f'(3) = \lim_{h \to 0} \frac{(3+h)^2 - 3^2}{h}$$

$$= \lim_{h \to 0} \frac{9 + 6h + h^2 - 9}{h}$$

$$\left[\frac{d \Delta p(s\phi)}{d\phi} \atop \frac{d \Delta M(s\phi)}{d\phi} \right] = \begin{bmatrix} \gamma & -\lambda \\ -\beta & 0 \end{bmatrix} \left[\Delta p(s,\omega) \atop \Delta M(s,\omega) \right]$$

$$= \lim_{h \to 0} \frac{6h + h^2}{h}$$

$$c_3 = c_{1+2} = \frac{1-1}{(1+1)(1+2)} c_1 = 0$$

$$= \lim_{h \to 0} (6 + h)$$

$$c_4 = c_{2+2} = \frac{2-1}{(2+1)(2+2)} c_2 = \frac{}{3}$$

$$= 6$$

$$c_5 = c_{3+2} = \frac{3-1}{(3+1)(3+2)} c_3 = \frac{}{4}$$

$$m) dx = \int (\log x)^2 dx = \frac{3}{2} \left\{ \frac{\pi^2}{12} + (\log 2)^2 \right\}$$

$$c_6 = c_{4+2} = \frac{4-1}{(4+1)(4+2)} c_4 = \frac{}{5}$$

$$n x + x^2 (\ln x)$$

$$c_7 = c_{5+2} = \frac{5-1}{(5+1)(5+2)} c_5 = \frac{}{6}$$

$$+ x^2 \left(\frac{1}{x} \right)$$

$$+ x$$

$$\frac{dP}{dt} = kP(1 - \frac{P}{n})$$

$$\frac{M}{P(M-P)} = \frac{A}{P} + \frac{B}{M-P}$$

$$\delta = \phi E - T$$

$$x + 1)$$

$$\frac{dP}{dP(1 - \frac{P}{n})} = kdt$$

$$M = (M-P)A + PB$$

$$P = M$$

$$P(w) = \int_{-\infty}^{\infty} P(x) e^{-2\pi x w} dx \frac{df}{d\omega}$$

$$M = MB$$

$$P(\frac{2v}{2t} + v \cdot \nabla v) = \nabla p + \nabla T + f$$

$$\int \frac{M dP}{P(M-P)} = \int k dt +$$

$$1 = B$$

$$P = D$$

$$H = \sum p(x) \log$$

$$\int (\frac{1}{P} + \frac{1}{M-P}) dP = kF + c$$

$$M - MA$$

$$\frac{1}{2} 6^2 S \frac{2v}{2.5} \cdot S \frac{2v}{35} \cdot \frac{2v}{2t} - v$$

$$A = 1$$

$$T((Q, q_n m)) = \sum_{n=1} [\frac{D}{m^2} S_i$$

$$= 0$$

$$= \frac{1}{3 \cdot 4} \frac{-c_0}{2}$$

$$= \frac{2}{4 \cdot 5} \cdot 0 = 0$$

$$= \frac{3}{5 \cdot 6} \frac{-c_0}{2 \cdot 3 \cdot 4}$$

$$= \frac{4}{6 \cdot 7} \cdot 0 = 0$$

$$f(x) = x^2 \ln x$$

$$f'(x) = (x^2)' \ln x + x^2 (\ln x)'$$

$$f'(x) = 2x \ln x + x^2 (\frac{1}{x})$$

$$f''(x) = 2x \ln x + x$$

$$f'(x) = x(2 \ln x + 1)$$

$$\delta = \phi E - T$$

$$P(w) = \int_{-\infty}^{\infty} P(x) e^{-2\pi x w} dx \frac{df}{d\omega}$$

$$i \frac{}{2t} \Psi = H \Psi$$

$$\sum_{n=1} \frac{\delta}{2} H_i^{M} c_s \frac{P}{a} c_0 D_z$$

$$P(\frac{2v}{2t} + v \cdot \nabla v) = \nabla p + \nabla T + f$$

$$\frac{Q(p-1)}{2p} H^M f_0 N$$

$$H = \sum p(x) \log p(x) \qquad F_0 N = \sum_{c=1} D_{i,\nu,d,(1+d)}$$

$$\frac{1}{2} 6^2 S \frac{2v}{}$$

CHANGE
IS THE ONLY
CONSTANT

THE WISDOM OF CALCULUS IN A MADCAP WORLD

BEN ORLIN

AUTHOR OF *MATH WITH BAD DRAWINGS*

BLACK DOG
& LEVENTHAL
PUBLISHERS
NEW YORK

Black Dog & Leventhal Publishers
Hachette Book Group
1290 Avenue of the Americas
New York, NY 10104

www.hachettebookgroup.com
www.blackdogandleventhal.com

First Edition: October 2019

Black Dog & Leventhal Publishers is an imprint of Perseus Books, LLC, a subsidiary of Hachette Book Group, Inc. The Black Dog & Leventhal Publishers name and logo are trademarks of Hachette Book Group, Inc.

The publisher is not responsible for websites (or their content) that are not owned by the publisher.

The Hachette Speakers Bureau provides a wide range of authors for speaking events. To find out more, go to www.HachetteSpeakersBureau.com or call (866) 376-6591.

Print book interior design by Headcase Design

LCCN: 2019930256

ISBNs: 978-0-316-50908-4 (hardcover); 978-0-316-50906-0 (ebook)

Printed in the United States of America

LSC-C

10 9 8 7 6 5 4 3 2 1

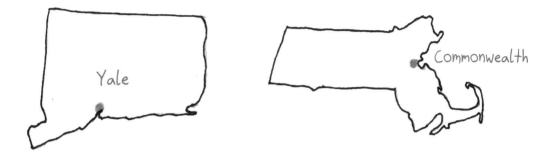

Yale

Commonwealth

FOR THE STUDENTS AND TEACHERS
AT ALL THE SCHOOLS I'VE CALLED HOME

Oakland Charter
High School

King Edward's School

There was a period of silence. After a while, he said,
"How did you get your ideas about God?"

"I was looking for God," I said. "I wasn't looking for mythology
or mysticism or magic. I didn't know whether there was a god
to find, but I wanted to know. God would have to be a power
that could not be defied by anyone or anything."

"Change?"

"Change, yes."

"But it's not a god. It's not a person or an intelligence
or even a thing. It's just...I don't know. An idea."

I smiled. Was that such a terrible criticism?

—OCTAVIA BUTLER,
PARABLE OF THE SOWER

CONTENTS

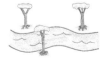

16. IN LITERARY CIRCLES,

in which calculus slices
a cucumber

17. WAR AND PEACE AND INTEGRALS,

in which calculus
revolutionizes history

18. RIEMANN CITY SKYLINE,

in which calculus becomes
an urban planner

19. A GREAT WORK OF SYNTHESIS,

in which calculus
hosts a dinner party

20. WHAT HAPPENS UNDER THE INTEGRAL SIGN STAYS UNDER THE INTEGRAL SIGN,

in which calculus expands
its toolkit

21. DISCARDING EXISTENCE WITH A FLICK OF HIS PEN,

in which calculus erases
68% of the known universe

INTRODUCTION

"What is," said the philosopher Parmenides, not quite a million days ago, "is uncreated and indestructible, alone, complete, immovable and without end." It's a bold philosophy. Parmenides permitted no divisions, no distinctions, no future, no past. "Nor was it ever, nor will it be," he explained; "for now it is, all at once, a continuous one." To Parmenides, the universe was like Los Angeles traffic: eternal, singular, and unchanging.

A million days later, it remains a very stupid idea.

C'mon, Parmenides. You can lull us with poetry and ply us with adjectives, but we're not dupes. A million days ago, there were no Buddhists, Christians, or Muslims, because Buddha, Jesus, and Muhammad had yet to be born. A million days ago, Italians did not eat tomato sauce, because "Italy" wasn't a concept and the closest tomatoes grew 6000 miles away. A million days ago, 50 or 100 million humans walked the Earth; now, that many people visit Disney-branded theme parks each year.

In fact, Parmenides, only two things were the same a million days ago as today: (1) the ubiquity of change, and (2) your philosophy's profound and irredeemable wrongness.

That's the last we'll hear of Parmenides in this book (although his savvier disciple Zeno will pop up later). Good riddance to toga-clad stoners, I say. For now, we jump ahead, past his wiser contemporary Heraclitus ("you can't step in the same river twice"), to arrive in the late 17th century, a mere 120,000 or 130,000 days ago. That's when a scientist named Isaac Newton and a polymath named Gottfried Leibniz birthed this book's protagonist. It was a fresh form of mathematics, a language of change, a stab at quantifying the flux and flow of this wobbling top called Earth.

Today, we call that math "calculus."

The first tool of calculus is the **derivative**. It's an instantaneous rate of change, telling us how something is evolving at a specific moment in time. Take, for example, the apple's velocity precisely as it strikes Newton's noggin. A second earlier, the fruit was moving a smidge slower; a second later, it will be moving in a different direction entirely, as will the history of physical science. But the derivative does not care about the second before, or the second after. It speaks only to *this moment*, to an infinitesimal sliver of time.

Calculus's second tool is the **integral**. It is the sum of infinite pieces, each infinitesimally small. Picture how a series of circles, each shadow-thin, can unify to create a solid object: a sphere. Or how a group of humans, each as tiny and negligible as an atom, can together constitute a whole civilization. Or how a series of moments, each of them zero seconds in itself, can amount to an hour, an eon, an eternity.

Each integral speaks to a totality, to something galactic, which the panoramic lens of our mathematics can somehow capture.

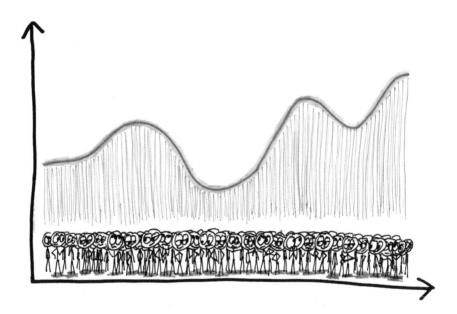

The derivative and integral have earned a lofty reputation as specialized technical tools. But I believe they can offer more. You and I are like little boats, knocked by waves, spun by whirlpools, thrown by rapids. The derivative and the integral, I hold, are pocket-sized philosophies: extendable oars for navigating this flood-swollen river of a world.

Hence, this book, and its attempts to distill wisdom from mathematics.

In the first half, **Moments**, we'll explore tales of the derivative. Each extracts an instant from the babbling stream of time. We'll consider a millimeter of the moon's orbit, a nibble of buttered toast, a dust particle's erratic leap, and a dog's split-second decision. If the derivative is a microscope, then each of these stories is a carefully chosen slide, a scene in miniature.

In the second half, **Eternities**, we'll call upon the integral and its power to unify infinite droplets into a single stream. We'll encounter a circle fashioned from tiny slivers, an army raised from myriad soldiers, a skyline built of anonymous buildings, and a cosmos heavy with a billion trillion stars. If the integral is a widescreen cinema, then each of these stories is a sweeping epic that you've *got* to see in theaters. The TV at home won't do it justice.

I want to be clear: the object in your hands won't "teach you calculus." It's not an orderly textbook, but an eclectic and humbly illustrated volume of folklore, written in nontechnical language for a casual reader. That reader may be a total stranger to calculus, or an intimate friend; I'm hopeful that the stories will bring a little mirth and insight either way.

This storybook is by no means complete—missing are the tales of Fermat's bending light, Newton's secret anagram, Dirac's impossible function, and so many others. But in an ever-changing world, no volume is ever exhaustive, no mythology ever finished. The river runs on.

BEN ORLIN

DECEMBER 2018

CHANGE
IS THE ONLY
CONSTANT

The moment of change is the only poem.

—ADRIENNE RICH

MOMENTS

MOMENT I.

Time claims another victim.

I.

THE FUGITIVE SUBSTANCE OF TIME

Jaromir Hladik has written several books, none to his satisfaction. One, he deems "a product of mere application." Another is "characterized by negligence, fatigue, and conjecture." A third attempted to refute a fallacy, but did so with arguments "not any less fallacious." I myself have only birthed books as flawless and sparkling as toothpaste commercials, but even so, I can empathize—especially with the little hypocrisy that gets Hladik through the day. "Like every writer," Jorge Luis Borges tells us, Hladik "measured the virtues of other writers by their performance, and asked that they measure him by what he conjectured or planned."

And what has Hladik planned? Oho! Hladik is glad you asked: It's a verse drama titled *The Enemies*, and it will be nothing less than his masterpiece. It will gild his legacy, cow his brother-in-law, even redeem "the fundamental meaning of his life"—if only he can clear the small hurdle of, you know, writing it.

Here I apologize, because our story takes a dark turn. Hladik—a Jew in Nazi-controlled Prague—is arrested by the gestapo. A perfunctory trial leads to a death sentence. On the eve of his execution, he prays to God:

> *If I exist at all, if I am not one of Your repetitions and errata, I exist as the author of* The Enemies. *In order to bring this drama, which may serve to justify me, to justify You, I need one more year. Grant me that year, You to whom belong the centuries and all time.*

The sleepless night passes, the execution day dawns, and then, just as the sergeant barks the final command to the firing squad, just as Hladik braces for death, just as all appears irretrievably lost...the universe freezes.

God has granted him a secret miracle. This single instant—with a raindrop rolling down his cheek and the fatal bullets still en route—has been enlarged, extended, dilated. The world is suspended, but his thoughts are not. Now Hladik can complete his drama, composing and polishing the stanzas entirely in his mind. The moment will endure for a year.

Here, on the cusp of a fate no one could envy, Hladik receives a gift that's the envy of all.

"The aim of every artist," William Faulkner once wrote, "is to arrest motion, which is life, by artificial means and hold it fixed." (Hladik himself is, of course, a work of fiction, by author Jorge Luis Borges.) *Tempus fluit*, wrote Isaac Newton: "time flows." *Tempus fugit*, declared sundials in the Middle Ages: "time flees." Although our purposes vary, all of us—artists, scientists, even those glib know-nothings we call "philosophers"—chase the same impossible prize. We want to grasp time, to hold the singular moment in our hands, the way that Hladik did.

Alas, time dodges and evades. Consider the famous "paradox of the arrow," from incurable Greek troll Zeno of Elea.

The idea: Picture an arrow flying through the air. Now, in your mind, freeze it at a single moment, like Hladik's firing squad. Is the arrow still moving? No, of course not—a freeze-frame is, by definition, frozen. In any given instant, the arrow is motionless. But if time is made of moments...and the arrow is in no moment ever moving...then how, exactly, can it move?

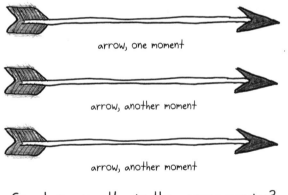

arrow, one moment

arrow, another moment

arrow, another moment

So when, exactly, is the arrow moving?

Philosophers in ancient China played similar mind games. "The dimensionless cannot be accumulated," one wrote. "Its size is a thousand miles." In the mathematical sense, a moment is *dimensionless*: It possesses no length, no duration. It's zero seconds long. But since two times zero is still zero, *two* moments will also amount to zero time. The same holds for 10 moments, or a thousand, or a million. In fact, *any* countable number of moments will still total up to zero seconds.

But if no supply of moments ever amounts to any time, then where do months and years and cricket matches come from? How can infinitesimal moments make up an infinite timeline?

Virginia Woolf noted that time "makes animals and vegetables bloom and fade with amazing punctuality." But it "has no such simple effect upon the mind of man. The mind of man, moreover, works with equal strangeness upon the body of time."

We chase the moment across history, mutilating time as we go. With hourglasses and candle clocks, we sliced days into hours. With pendulums and escapements, we carved hours into minutes (the etymology: "a *minute* fraction of an hour"), and thence into seconds (as in "a second order" of tiny; a minute fraction of a minute). From there we decomposed time into milliseconds (half a flap of a fly's wings), microseconds (a fancy strobe light's flash), and nanoseconds (each enough time for light to travel a foot), not to mention pico-, femto-, atto-, zepto-, and yoctoseconds. After that the names peter out, presumably because Dr. Seuss ran out of ideas, but still we slice on. Eventually, eternity crumbles into units of "Planck time," about a billionth of a trillionth of a yoctosecond, or just enough for light to travel $\frac{1}{100,000,000,000,000,000}$ of the way across a proton. No instrument can reach beyond this ultimate in brevity: physicists insist that it's the smallest meaningful unit of time in the universe, as far as we understand (or, like me, fail to understand).

LENGTH	SECONDS	SIGNIFICANCE
1 minute	60	Longest recorded gap between super-hero movie releases
1 second	Uh...1	Length of a sneeze, or 0.1% of a kilosneeze
1 millisecond	$\dfrac{1}{1000}$	Average human attention span
1 microsecond	$\dfrac{1}{1,000,000}$	Length at which video buffering becomes intolerable
1 nanosecond	$\dfrac{1}{1,000,000,000}$	Time required for a dog to decide it doesn't trust me
1 Planck time	$\dfrac{1}{10^{43}}$	Time after which I lose the thread when physicists discuss quantum effects, such as the Planck time
1 moment	Zero	?!?!?!?!?!?!?!

Where, oh where, is "the moment"? Is it somewhere past Planck? If we can neither gather moments into intervals, nor break intervals into moments, then what even *are* these invisible, indivisible things? As I write this book in the tick-tock world of ordinary time, in what effervescent non-world is Hladik writing his?

In the 11th century, mathematics first articulated a tentative answer. While European mathematics was pulling out its hair trying to calculate the date of Easter, Indian astronomy was busy predicting eclipses. This required pinpoint precision. Astronomers began throwing around units of time so brief that it would be almost a millennium before any timepiece could possibly measure them. One *truti* amounted to less than 1/30,000 of a second.

These practically infinitesimal slivers of time paved the road to a concept called *tatkalika-gati*: instantaneous motion. How fast, and in what direction, is the moon moving *right at this exact moment*?

And then, what about *this* moment?

And what about *now*?

And *now*?

These days, *tatkalika-gati* goes by a more pedestrian name: "the derivative."

Consider a speeding bicycle. The derivative measures how fast its position is changing—i.e., the bike's velocity in a given moment. In the graph below, it's the steepness of the curve. A steeper curve signifies a faster bike, and thus a greater derivative.

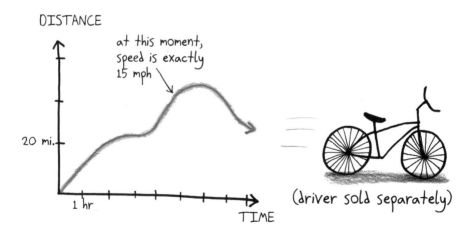

DISTANCE

at this moment, speed is exactly 15 mph

20 mi.

1 hr

TIME

(driver sold separately)

Of course, in any given moment, the bicycle is like Zeno's arrow: motionless. Thus, we can't calculate derivatives via freeze-frame. Instead, we work by zooming in. First, determine the bike's speed over a 10-second interval; then, try a 1-second interval; then, a 0.1-second interval, and a 0.01-second interval, and a 0.001-second interval...

In this sly manner, we sneak up on the instant, drawing closer and closer and closer until a pattern becomes clear.

START	END	SPEED
12:00:00	12:00:10	39 mph
12:00:00	12:00:01	39.91 mph
12:00:00	12:00:00.1	39.98 mph
12:00:00	12:00:00.01	39.997 mph
At precisely noon...		*40 mph*

For another example, take a fizzing reaction, as two chemicals wed their little chemical parts to form a new baby chemical. The derivative measures how fast the product's concentration is growing—i.e., the rate of the reaction in a given moment.

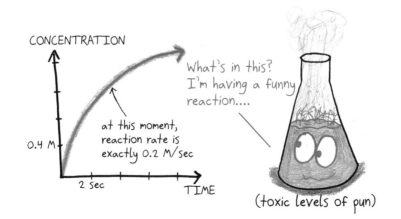

Or consider an island overrun with rabbits. The derivative measures how fast the population's size is changing—i.e., its rate of growth in a given moment. (For this graph, we must briefly entertain the fiction of "fractional rabbits," but if your suspension of disbelief has come this far, I trust it to survive any challenge.)

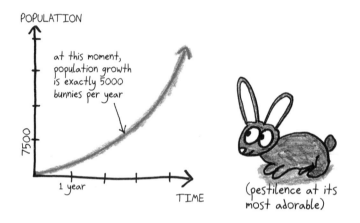

at this moment, population growth is exactly 5000 bunnies per year

(pestilence at its most adorable)

This bread-and-butter mathematical concept is oddly like a poet's fancy. The derivative is "instantaneous change": movement captured in a moment, like lightning in a bottle. It's the repudiation of Zeno, who said that nothing can happen in a single instant, and the vindication of Hladik, who believed that anything can.

By now, perhaps you've guessed the end of Hladik's tale. For 12 months, he composes his play. He writes not "for posterity," Borges tells us, "nor even for God, of whose literary preferences he possessed scant knowledge." Instead, he writes for himself. He writes to satisfy what Thomas Wolfe considered the perpetual itch of the artist:

> to fix eternally in the patterns of an indestructible form a single moment of man's living, a single moment of life's beauty, passion, and unutterable eloquence, that passes, flames and goes, slipping for ever through our fingers with time's sanded drop, flowing for ever from our desperate grasp even as a river flows and never can be held.

Hladik has held the river. It does not matter that no one will read *The Enemies*, or that the bullets will, in no time at all, resume the course. It matters only that he has composed his book, that it will now live forever, in this single moment, which is its own kind of eternity. ∎

MOMENT II.

Isaac Newton decides the moon is an apple, and vice versa.

II.

THE EVER-FALLING MOON

Isaac Newton was a curious child. I mean "curious" as in "always questing for knowledge," and also as in "utterly weird." According to one story, he would become so engrossed in reading that his pet cat grew fat off of his untouched meals. Or consider his first exploration of optics. Ever meet a kid so curious he'd risk his own sight for a glimpse of the truth? He writes in his journal: "I took a bodkin"—that's a thick, blunt needle—"& put it betwixt my eye & [the] bone as neare to [the] backside of my eye as I could: & pressing my eye…there appeared severall white, darke & coloured circles."

It's a shame, but today we rarely remember Newton as a self-mutilating owner of obese housecats. Instead, we remember him as a guy who got hit on the head by fruit.

Actually, the noggin impact was a later embellishment. As Sir Isaac himself told the tale, all it took was a glimpse of a falling apple to set the clockwork of his mind into historic motion. "As he sat alone in a garden," recalled Henry Pemberton, a personal friend of Newton's, "he fell into a speculation on the power of gravity." The apple's fall prompted him to reflect that no matter how high we rise—rooftops, treetops, mountaintops—gravity does not diminish. It is, to repurpose a phrase from Albert Einstein, "spooky action at a distance." The matter of Earth seems to attract the matter of objects, no matter how far removed they are.

The curious young man probed further. (No bodkins this time; just thoughts.) What if gravity reaches *beyond* the mountaintops? What if its pull extends outward farther than we've guessed?

What if it goes all the way to the moon?

Aristotle would never have believed it. The stars obey perfect patterns, symphonic cycles, like my wife's family organizing a dinner party. Life on Earth is anarchy, a mud splatter, like me organizing a dinner party. How could the two realms possibly follow the same laws? What eye-gouging madman would dare unify the terrestrial and the celestial?

Well, in the spring of 1666, that madman was 23 years old, relaxing in the shade of his mother's garden. He watched an apple fall, and then, by some inspired stroke, he imagined a second falling apple, this one as distant as the moon. One small step for a McIntosh, one giant leap for fruitkind.

He knew the distance, roughly: if Earth's surface sits one unit from its center, then the moon is 60 units away.

Total Distance: 60 times Earth's radius

384,000 Km

6400 Km

At such a tremendous remove, how might gravity act?

Even the loftiest mountains offer no clue. Compared to the moon, the peak of Everest sits practically on the skin of the Earth, only a cosmic hair's breadth away. But let's suppose—in an enormous and only slightly ahistorical leap—that gravity decays at greater distances. The farther you go, the weaker its force. I'm referring to Newton's famous "inverse square law":

At twice the distance, there's 1/4 the gravity.

At triple the distance, 1/9 the gravity.

At ten times the distance, just 1/100 the gravity.

Our brave spacefaring apple, 60 times farther away from Earth's core than its timid orchard-dwelling cousins, would experience just 1/3600 the gravity. If you've never divided something by 3600, let me editorialize: it makes things a lot smaller.

Drop an apple near the surface of the Earth, and it falls 4.9 meters in the first second. That's about the height of a second-story window.

Drop our astro-apple from moon height, and in the first second it descends just over 1 millimeter. That's the thickness of a nice credit card.

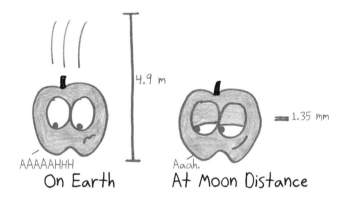

Back then, the explanation of the moon's orbit remained an open mystery. René Descartes's vortex theory—that all heavenly objects are swept along in their paths by swirling bands of particles, like bath toys circling a drain—reigned as the favored theory. But this was a time of change: Newton's *annus mirabilis*, his "miracle year" (which "miraculously" lasted more like 18 months). During this solitary stretch at his mother's cottage in Woolsthorpe, England, waiting out the plague that was ravaging London, Newton developed the ideas that would launch modern math and science. He articulated his laws of motion, unlocked the optical secrets of the prism, managed to keep his eyes free of household objects, and discovered calculus.

Along the way, he dethroned Descartes's vortices with the toss of an apple.

As Newton's predecessor and soul brother Galileo knew, horizontal motion does not affect vertical motion. Drop one apple; launch an identical apple sideways; and they'll hit the ground at the same moment. Sure, their horizontal trajectories diverge, but their vertical motions obey the same dictatorial force: gravity.

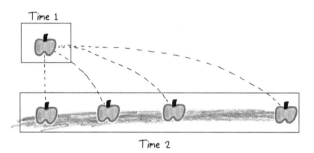

Now, take your apples to a very high mountaintop, and throw them with superhuman speed. Congratulations: you've stepped inside a celebrated diagram from Newton's masterwork (the *Principia*), illustrating the peculiar physics of high-speed falling.

Here, thanks to the planet's curvature, our tidy vertical/horizontal distinction evaporates. One moment's "horizontal" is the next moment's "vertical." The stronger the throw, the more prolonged the fall.

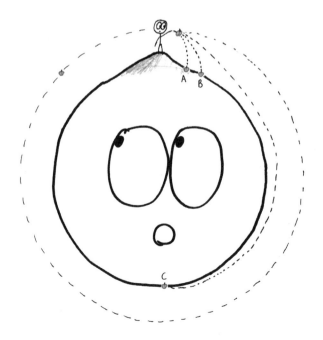

Throw the apple hard—like, major-league-pitcher hard—and it will travel a little distance before falling to earth. It may reach point A or B.

Throw the apple *really* hard—like, Red-Sox-pitcher-throwing-at-an-uppity-Yankees-player hard—and its horizontal motion now leads it away from the planet, prolonging the fall. Perhaps it travels all the way to C.

Throw the apple *stupendously* hard—like, Henry-Rowengartner-on-steroids hard—and it flies away from Earth so fast that each moment's falling merely restores the apple to its original height. The apple can thereby fall forever.

An orbit is just a perpetual fall, with no Cartesian vortices required.

How would this work with our intrepid moon-apple? Well, this is calculus, so consider a virtually infinitesimal moment: a single second of travel. Over such a brief interval, the curved arc of the orbit might as well be a straight line.

Here we indicate the distance that the apple would fall if left to gravity's devices alone.

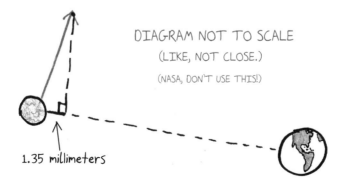

Now what? Newton's next move is a nifty geometric argument. We've created a little right-angled triangle. We want to know its hypotenuse (i.e., its longest side). So we embed it in a larger triangle that shares the same proportions:

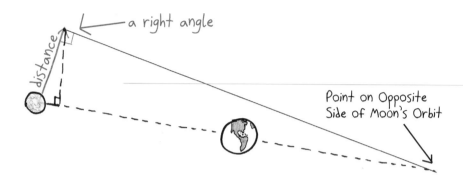

Because the triangles are the same shape, their sides relate by the same ratio:

$$\frac{1.35 \text{ mm}}{\text{distance}} = \frac{\text{distance}}{781,542 \text{ Km}}$$

Solving this equation yields the following solution:

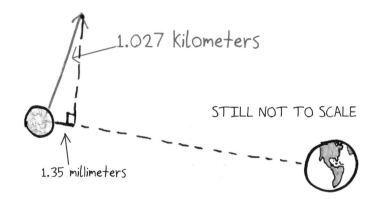

Our apple descends, you may recall, at the gentle rate of 1 millimeter per second—about 3% the ground speed of a sloth. And yet, to keep the fruit in orbit, we must fling it sideways at a speed of 1 *kilometer* per second—about triple the speed of sound.

This strikes me as simple, extraordinary—and, on its face, implausible. The moon, falling like a flung apple? Really, Sir Isaac? Can you confirm this wackadoodle thought experiment with any—oh, what's it called—*evidence*?

Well, consider the time our moon-apple needs to circle the Earth. At such an enormous distance, it will have to traverse a path 2.5 million kilometers in circumference. Moving just over 1 kilometer per second, how long will this take?

$$2,390,737 \text{ seconds} = 27.7 \text{ days}$$

Ha, look at that! Our computation matches—to within less than 0.7%—the length of the moon's actual orbit. This number gives stunning confirmation of Newton's theory: the moon really is falling like a giant Red Delicious (and is about as rewarding to eat). As biographer James Gleick puts it:

> *The apple was nothing in itself. It was half of a couple—the moon's impish twin... Apple and moon were a coincidence, a generalization, a leap across scales, from close to far and from ordinary to immense.*

Short of claiming that Newton invented friendship and the color purple, it's hard to overstate the impact of Sir Isaac's theory. It identified a single universal force to govern the earthly and heavenly realms, birthing a modern vision of reality: the mechanical universe, a clockwork cosmos that obeys definite and unbreakable laws as it evolves from moment to moment.

French scholar Pierre-Simon Laplace put it like this: Imagine a vast intellect that knows the location of every object and the strength of every force. Such a mind would know, well, everything. "Nothing would be uncertain," said Laplace, "and the future just like the past would be present before its eyes."

All the world's a differential equation, and the men and women are merely variables.

Not everyone celebrated the Newtonian vision. The poet William Blake, not one to mince words, declared, "Science is the tree of death." Writer Alan Moore elaborates: "For Blake, the boundaries of Newton's thought were the cold, stone parameters of an internal dungeon to which all humanity had been condemned."

Heavy stuff.

Still, Newton had literary defenders by the hordes. Beating out runners-up Alexander Pope ("Nature and Nature's laws lay hid in Night: / God said, 'Let Newton be!'—and all was light") and William Wordsworth ("The marble index of a mind for ever / Voyaging through strange seas of Thought, alone"), Newton's most fervent advocate was the philosopher and science fanboy Voltaire, who called him "the creative spirit," "our Christopher Columbus," and (perhaps overdoing it) "the God to whom I sacrifice." To Voltaire, we owe one of history's more poetical descriptions of calculus— "the art of numbering and measuring exactly a thing whose existence cannot be conceived"—as well as the popularity of the apple story, which he placed at the center of Newton's intellectual journey.

With this mist of myth hanging around it, how far can we trust the apple tale?

If you like to bet on winners
then you better call your bookie;
I'm a blur, a Sir, a legend,
and a chewy fig cookie.

I'm a one-man show of genius
and I don't need no rehearsal,
'cause my fame is like
my law of gravitation:
universal.

"The story was certainly true," says Keith Moore, head of archives at the Royal Society, "but let's say it got better with the telling." Newton himself played up the anecdote, perhaps at the expense of a more honest account of the fits and starts by which science really progresses. After all, he'd spend another 15 years refining his theories, drawing on work from Galileo, Euclid, Descartes, Wallis, Hooke, Huygens, and countless others. Theories do not just blink into existence; they have roots. They grow. That moment in the garden did not sprout a fully formed understanding of gravity—it merely gave us the first sunlit glimpse of the seedling. ■

MOMENT III.
With apologies to Gerard Manley Hopkins.

III.

THE FLEETING JOYS
OF BUTTERED TOAST

When I moved to England and first stepped into the 462-year-old private school where I'd be teaching, I could not believe my luck. Every morning at break time, teachers retired to the faculty lounge to enjoy pots of tea and trays of toast. The concepts of "faculty lounge" and "break time" had already elevated this workplace above my prior one. But a daily morning feast, a living daydream of Hogwarts? "I'll never get used to this!" I told my new colleagues.

I got used to it.

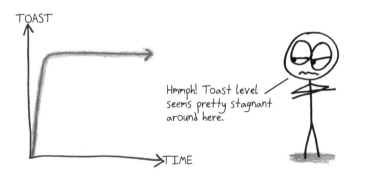

Psychologists call this process *habituation*. It means that I've got vision like a dinosaur's: keenly trained on whatever moves, I overlook anything that stands still, even if it comes buttered. Perhaps evolutionary psychology can explain the phenomenon, or perhaps I'm an ungrateful louse, but either way, you can frame habituation mathematically. We grow accustomed to the function, whatever its height. Before long, it takes a derivative—a nonzero rate of change—to draw our attention. Only a newer newness can catch our eye.

One day, cradling a fresh mug of tea and munching a piece of wheat toast (*ugh—I thought I grabbed white*), I plopped onto a sofa next to my friend James, an English teacher. "How's it going?" I greeted him.

James took this placeholder question like he takes everything: in complete and utter earnest.

"I'm happy this week," he reflected. "Some things are still hard, but they've been getting better."

Evidently, I'm a math teacher first and a human being second, because this is how I responded to my friend's moment of openness: "So, your happiness function is at a middle sort of value, but the first derivative is positive."

James could have slapped the toast from my hand, dumped his tea over my head, and screamed, *Friendship annulled!* Instead, he smiled, leaned in, and said—I swear this is a true story—"That's fascinating. Explain to me what that means."

"Well," I lectured, "picture a graph of your happiness over time. Yours is at a medium height. But as of this moment, it's rising—that's a positive derivative."

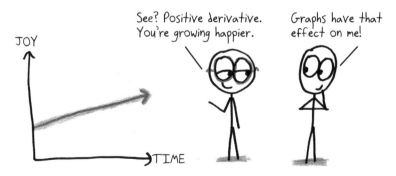

"I see," he said. "So a negative derivative means it's getting worse?"

"Well," I hedged, "sort of." I was showcasing the pedantry for which mathematicians are so beloved. (Or is it pronounced "reviled"?) "Negative derivative means the value is decreasing. For some functions—like personal debt, or physical pain—you'd *want* a negative derivative. But in the case of happiness, yes, it's a bad thing."

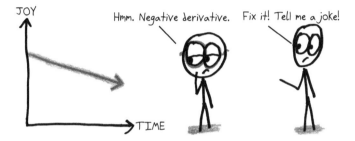

It was a rather unconventional first lesson in differential calculus. Most students encounter these ideas not through the fuzzy psychology of the "happiness" function, but through the crisp physics of "position." For example, let p represent the location of a cyclist along a bike path. The starting point is $p = 0$; half a mile later, $p = 2640$ feet.

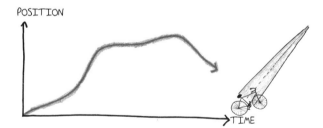

What's the derivative here? Well, it's how fast p is changing at a particular moment in time. We call it p' (pronounced "p prime"), or (more evocatively) "velocity."

A large value for p'—say, 44 feet per second—means the position is changing quickly; velocity is high. A small value—say, 2 feet per second—signals a low velocity. If p' is zero, then the position is not changing at all; the bicycle is stationary. And if p' is negative, then we're moving backward along the path; the bicyclist has reversed direction.

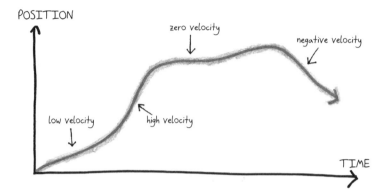

From our original graph (specifying each moment's *position*) we can "derive" a whole new one, specifying each moment's *velocity*. That's where the word "derivative" originates; this act of deriving, or derivation.

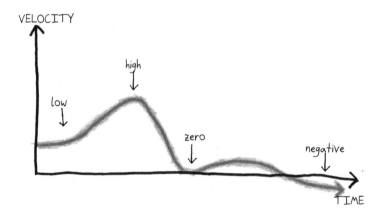

James, bless him, was soaking up the calculus like it was some kind of alien poetry. As an English teacher, he is a professional investigator of language and its power to capture human experience. Here, in the dry dialect of derivatives, he seemed to have found an unlikely kinship, a clunky form of literature in translation.

"And then there's the second derivative," I said.

James nodded with severity. "Tell me."

"It's the derivative of your derivative—so it tells how your rate of change is changing."

James frowned, for the understandable reason that I was speaking nonsense.

I tried again. "The derivative is your rate of improvement. The second derivative asks: Are you improving faster and faster? Or is the improvement slowing down?"

"Hmm." James rubbed his chin. "For me, I'd say faster and faster. So the second derivative is...positive, right?"

"Yes!"

"And if the improvement were slowing down," he continued, "then the first derivative is still positive, but the second is negative."

"Yes."

"I like this," said James. "I should teach this to all of my friends. Then, when they ask me for an update, I can convey my precise emotional state just by reciting a few numbers."

"So, like, h positive, h prime negative, h double prime positive?"

"Ooh, let me see." James heard in my statement a linguistic puzzle, a concise and artless form of memoir. "That means...I'm happy...and I'm getting less happy...but the decline in my happiness is slowing down?"

"That's right."

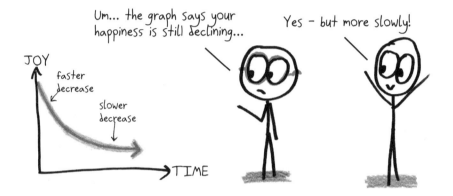

For capturing delicate shades of emotion, this language may seem stilted or crude, like bellowing *"Human happy!"* and *"Human sad!"* But like all derivatives, it's a kind of physical metaphor—an analogy to motion through space.

As we saw with our bicycle, position's derivative is velocity. And velocity's derivative? That's *acceleration*. (It's also called p'', or "p double prime": an oxymoronic name, since "prime" means "first.")

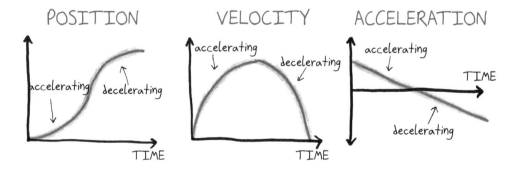

Derivatives and second derivatives give distinct information. To see the difference, picture a rocket moments after takeoff, the astronauts' faces pressed downward like jelly. The velocity is still low, but changing quickly, so the acceleration is high.

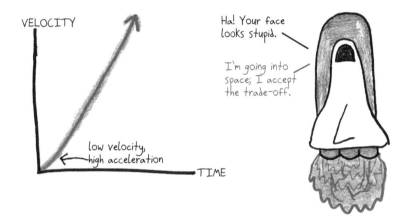

The reverse can also happen. A cruising airplane has a high velocity, but it's a steady and unchanging velocity, so the acceleration is zero.

(As these examples illustrate, speed doesn't affect our bodies much. What makes a biomechanical difference—what pressures, nauseates, bewilders, and excites us—is acceleration, because it corresponds to force.)

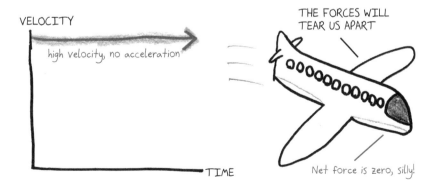

"Poetry begins in trivial metaphors," Robert Frost once wrote, "pretty metaphors, 'grace' metaphors, and goes on to the profoundest thinking that we have." I'm not sure Frost would have found much poetry in derivatives—they are hopelessly direct, saying one thing only, and with lamentable precision—but the soil here is fertile with metaphor. As velocity tells us the change in position, so does acceleration tell us the change in velocity—and so does the appropriate derivative tell us the change in happiness.

James, no minor master of metaphor himself, knew the next question to ask: "What about the *third* derivative?"

In physics, the third derivative (p''', or "p triple prime") is called *jerk*. It refers to a change in acceleration—which is to say, a change in the forces acting on a body. Think of the moment a car slams on its brakes, or the precise instant that a rocket blasts off, or the microsecond when a fist strikes a face. A new force arrives. The acceleration changes.

I've never taught jerk, except as a novelty. Three derivatives are an awful lot. "Certainly he who can digest a second or third Fluxion," wrote the 18th-century philosopher George Berkeley, using Newton's word for derivatives, "need not, methinks, be squeamish about any Point in Divinity."

"It's kind of hard to understand," I warned James. "The physical interpretation is pretty subtle."

But in the five preceding minutes, I had won a convert, which is to say, a zealot. "Don't give up now!" James cried. "The third derivative is simple: it's the change in the change in the change in my happiness." His volume began to rise; colleagues looked over with concern. "In fact, I must give *all* of the derivatives! An infinite catalogue of numbers describing how my happiness is changing, and how the change is changing, and how the change of the change of the change is changing...Then my friends could know exactly how I feel without a word spoken."

"True," I said. "In fact, if they know precisely how your happiness is evolving in this moment—the whole infinite chain of derivatives—then they can forecast your emotional state indefinitely far into the future. With enough derivatives, they can infer the course of your happiness for your entire life."

"Even better!" James laughed maniacally and clasped his hands. "I'll never need to speak to my friends again!"

I grew worried. "Wouldn't that, in itself, have a negative effect on your happiness?"

James waved the objection off. "I'll just build that into the derivatives. They'll know."

That's when the bell rang. Even in a teacher's paradise, you've got to deliver lessons from time to time. I left my tea mug on the counter, dashing off to my classroom. I'd like to think I mumbled a thank-you to Sarah, the woman who set out the toast and washed our dishes, but I know that, habituated monster that I am, there were days when I forgot. ∎

Sing, Muse, of the Infinitesimal,
how it battled and baffled
all who touched it,
until a symbol soothed its rage,
and the child Calculus was born.

MOMENT IV.

Gottfried Leibniz tells his epic tale.

IV.

THE UNIVERSAL
LANGUAGE

I love coining mathematical words. At least, I love trying. The brutal truth is that "canceltharsis" (the gratification of terms canceling out) and "algebrage" (the incandescent anger of losing hours to a small algebraic error) haven't really caught on yet. Alas, this is yet another way in which Gottfried Leibniz's achievements exceed mine, because he gave to the mathematical lexicon words such as:

- Constant, for a quantity that doesn't change;
- Variable, for a quantity that does;
- Function, for a rule associating inputs to outputs;
- Derivative, for an instantaneous rate of change; and
- Calculus, for a system of calculation, like the one he developed.

Throw in the symbols he popularized without inventing (e.g., $\cong$ for congruence, = for proportions, and the use of parentheses for grouping), and it becomes clear that writing mathematics in the 21st century means walking a trail Leibniz blazed in the 17th. Even so, all of these are just footnotes to his greatest notational contribution of all.

The letter d.

It sounds awfully simple. More Sesame Street than Harvard Yard. "All Leibniz did was put d in front of x," joked the legendary mathematician Sir Michael Atiyah in 2017. "Apparently, you can become famous that way."

To be fair, notational breakthroughs always feel obvious in hindsight. How often do you thank Robert Recorde, inventor of the = sign, for sparing us the endless repetition of "is equal to"? The purpose of mathematical symbols is to let us project our thoughts onto paper. Well-chosen ones feel so natural that you forget how artificial the whole process is. Make no mistake: mathematical symbolism is a technological feat, the extension of the brain by other means, as eerie and profound as a robotic limb.

And no one in history has crafted symbols with the same vivid clarity as Leibniz. "I suspect Leibniz's successes in mathematics," reflects computer scientist Stephen Wolfram, "were in no small part due to the effort he put into notation."

Born in 1646, just a few years after his calculus coparent Newton, Leibniz enjoyed an eclectic career. A philosopher, a socialite, and, as portraits reveal, a sentient vehicle for giant wigs, he counted "discovering calculus" as just one item on a towering CV. He was Europe's foremost expert on geology, on China, on difficult cases in the law—and, more generally, Europe's foremost expert. One royal employer referred to him with a rueful sigh as "my living dictionary." Over his lifetime, he wrote 15,000 letters to more than 1000 correspondents.

Leibniz cared about his readers. Unlike Newton, who deliberately wrote the *Principia* in a difficult style ("to avoid being baited by the little Smatterers in Mathematicks"), Leibniz valued clear communication. And so, as he developed the concepts of calculus, he made sure to clothe them in smart and well-fitted symbols.

Symbols like *d*.

In mathematics, Δ (the Greek letter "delta") signifies change. Consider a ripped-from-the-headlines example, which occurred this morning, and would have been unheard of just six months ago: I went for a run.

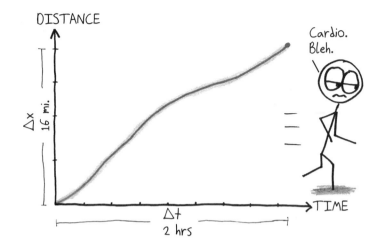

If x is my distance from home, then Δx is the change in that distance over a certain span. Let's say 16 miles (because this is my book, and I can lie if I wish).

Now, if t is the time, then Δt will be the time elapsed during my run. Let's say two hours (because, hey, it simplifies the arithmetic to make me a speed demon).

What was my speed? Well, to compute any rate of change, we divide. Here, it's Δx divided by Δt, which gives 8 miles per hour.

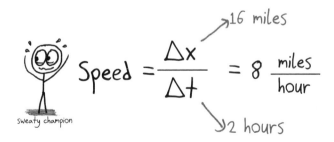

Now, what about my speed at precisely 1:00 p.m.? A derivative, you may recall, is an *instantaneous* rate of change. It does not analyze a leisurely span of two hours. It zooms in on a single moment, a freeze-frame.

But that poses a problem. During this infinitesimal span, no time elapses, and I cover no distance. Δx and Δt are both zero. And 0/0 does not yield a very illuminating answer.

Enter Leibniz's retooled notation. Instead of Δx and Δt, we consider dx and dt: infinitesimal increments of position and time.

Hence, his notation for the derivative: $\frac{dx}{dt}$.

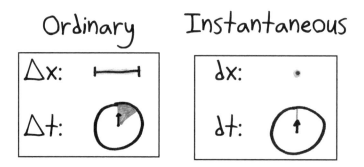

There's deception afoot: dx and dt aren't really numbers, and you can't really divide them. The notation isn't literal; it's more like an analogy, or a magician's sleight of hand. But that's exactly what makes the symbolism so powerful. Harvard mathematician Barry Mazur compares Leibniz's derivative to a semipictographic character from a language like Chinese or Japanese: not a mere arbitrary mark, but a tiny, evocative illustration of the concept's essence. He counts it among his "favorite pieces of mathematical terminology" for precisely this reason: it is "visually self-explanatory."

I must confess. As a student, I preferred the Newton-influenced' notation (which we met in Chapter 3). To me, the $\frac{dx}{dt}$ business felt cluttered, complicated, and worst of all, booby-trapped: a fraction that's not really a fraction.

But with time, I came to appreciate the secret power of Leibniz's d's: their enormous flexibility. Whereas the primes presume a single input variable (often time), Leibniz's symbols reach far wider. They allow us to choreograph huge casts of variables in complex ballets.

To see, let us step into the economics classroom. Or better yet, the boardroom of a toy company.

You and I make teddy bears, selling a certain quantity (q) at a certain price (p). What will happen if we raise the price by a tiny increment? In general, we'll sell fewer teddies, but the precise answer is a derivative: $\frac{dq}{dp}$. That's quantity's instantaneous rate of change with respect to price.

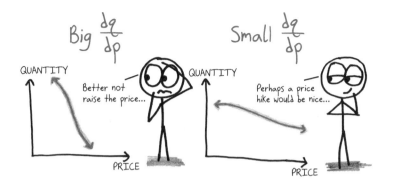

Still, q depends on more than p. Perhaps we advertise, spending a dollars on TV commercials. In that case, $\frac{dq}{da}$ spells out the marginal impact an extra advertising dollar has on sales.

Then again, if we advertise more, we may need to raise the price. That means considering $\frac{dp}{da}$: how the price we charge depends on changes to the advertising budget.

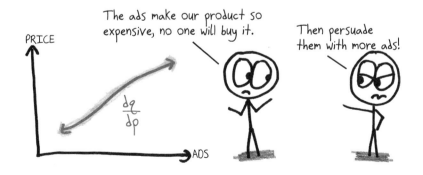

We can even flip our derivatives upside down. What about $\frac{dp}{dq}$? This tells us how the *price* will respond to an infinitesimal change in the *quantity*.

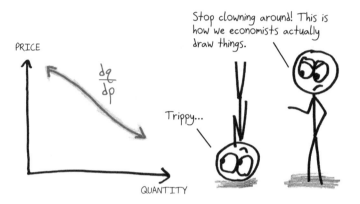

Can prime notation juggle such diverse derivatives? Please! Only Leibniz's nimble d's can manage the task with such grace and finesse. And that makes Leibniz's the perfect language for discussing calculus's deepest application: the art of optimization.

I don't know about you, but I'm not in the teddy bear business to make friends. I'm not even in it to make plush predators that weaken children's healthy and natural fear of bears. I'm here to make money, and as such, there is only one output variable that matters to me: *profit*.

objects with no intrinsic value

To maximize profit, we don't want to set the price too low. Say a teddy bear costs $5 to make; in that case, selling it for $5 is a charity, not a business. And $5.01 is scarcely better—sure, we'll sell lots of teddies at this discount rate, but even if we sell a million, we'll walk away with just $10,000.

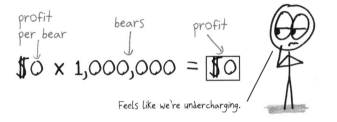

profit per bear · bears · profit

$$\$0 \times 1{,}000{,}000 = \boxed{\$0}$$

Feels like we're undercharging.

On the other hand, we don't want to set the price too *high*. If we charge $5000 per bear, then perhaps some naïve billionaire will buy one. Perhaps not. Either way, we'll sell too few to turn any substantial profit.

Now I fear we've overcompensated...

profit per bear · bears · profit

$$\$5000 \times 0 = \boxed{\$0}$$

What we need is a derivative. If we raise our price by some infinitesimal increment, then how will that affect our profit?

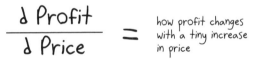

$$\frac{\partial\ Profit}{\partial\ Price} =$$

how profit changes with a tiny increase in price

Positive derivative? That means that jacking up the price would boost profits. In other words, we're undercharging.

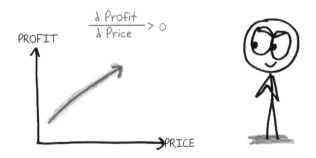

$$\frac{\partial\ Profit}{\partial\ Price} > 0$$

PROFIT

PRICE

Negative derivative? That means that *lowering* the price would boost profits, by reeling in more customers. In other words, we're overcharging.

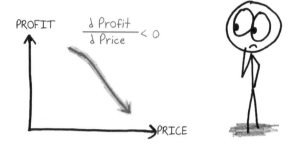

We want to find a special moment: the price at which the derivative is precisely zero.

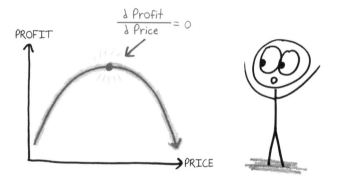

A *maximum* is a moment of transition: when your derivative switches from positive to negative. Meanwhile, a *minimum* is just the opposite: a moment when the derivative switches from negative to positive. The logic is actually quite plain: Go until things stop getting better, and are about to start getting worse. That's the best you can do.

We've defined a maximum not by its *global* properties ("it's the highest of all the points"), but by its *local* ones. Look a hair to the left; the graph slopes up. Look a hair to the right; the graph slopes down. Look right at the point, and the derivative is zero. This defines a maximum on the basis of a microscopic analysis. It's a nifty trick, like identifying a mountaintop from a soil sample.

The first publication in the history of calculus was Leibniz's 1684 paper *Nova Methodus pro Maximis et Minimis* ("New Method for Maximums and Minimums"). "Nothing takes place in the world," the mathematician Leonhard Euler once said, "whose meaning is not that of some maximum or minimum."

When Leibniz was in his early twenties, he decided to join an exclusive society of alchemists. (Hey, it was the 1660s; everybody was doing it.) To prove his alchemical bona fides, Leibniz compiled a list of buzzwords, from which he stitched together a lengthy, impressive, and quite nonsensical application letter. It worked; the dazzled alchemists elected him their secretary. But—surprise, surprise—Leibniz saw through the BS. He left within months, later denouncing the group as a "gold-making fraternity."

To me, this is classic Leibniz. First, you master the language. Then the truth, whatever it is, will emerge. Less than a decade after memorizing the gibberish vocabulary of alchemy, that brash young man would devise mathematical vocabulary used by millions to this day.

Did he ever turn lead into gold? Nope. He did one better: turning a lowercase d into a timeless language of the moment. ∎

MOMENT V.
Mark Twain teaches math.

V.

WHEN THE MISSISSIPPI RIVER RAN A MILLION MILES LONG

In the opening pages of *Life on the Mississippi*, Mark Twain gives the people what they crave: statistics. The Mississippi River's length: 4300 miles. Its basin: 1,250,000 square miles. Its annual deposits: 406 million tons. "This mud," Twain calculates, "would make a mass a mile square and two hundred and forty-one feet high." It's all very empirical, and perhaps—for a writer whose books are alternately hailed as hilarious and banned as blasphemous—a trifle dry.

But worry not, Twain fans! As the man himself said: "Get your facts first, and then you can distort them as much as you please." A virtuosic embellisher like Twain can weave tall tales out of anything—even numbers. To wit:

> *These dry details are of importance in one particular. They give me an opportunity of introducing one of the Mississippi's oddest peculiarities,—that of shortening its length from time to time.*

Like all old rivers, the Mississippi meanders through lazy, looping turns. In one stretch, it weaves 1300 miles to arrive a mere 675 from where it began. And, from time to time, the river will surge across a narrow stretch of land, short-circuiting a meander. "More than once," says Twain, "it has shortened itself thirty miles at a single jump!" In the two centuries prior to Twain's book, the river's lower length, between Cairo, Illinois, and New Orleans, Louisiana, dropped from 1215 miles to 1180 to 1040 to 973.

Here, the tale-teller takes over:

Geology never had such a chance, nor such exact data to argue from!...Please observe:—

In the space of one hundred and seventy-six years the Lower Mississippi has shortened itself two hundred and forty-two miles. That is an average of a trifle over one mile and a third per year. Therefore, any calm person, who is not blind or idiotic, can see that in the Old Oolitic Silurian Period, just a million years ago next November, the Lower Mississippi River was upwards of one million three hundred thousand miles long, and stuck out over the Gulf of Mexico like a fishing-rod. And by the same token any person can see that seven hundred and forty-two years from now the Lower Mississippi will be only a mile and three-quarters long, and Cairo and New Orleans will have joined their streets together, and be plodding comfortably along under a single mayor and a mutual board of aldermen. There is something fascinating about science. One gets such wholesale returns of conjecture out of such a trifling investment of fact.

Now, is Twain merely playing a silly game of arithmetic? Not at all! His is a profound game of geometry. It's the elemental geometry at the heart of calculus, the geometry that makes derivatives both possible and useful: the ubiquitous geometry of the straight line.

Please observe:—

We can create a graph, showing the length of the Lower Mississippi (from Cairo to New Orleans) at different years throughout history:

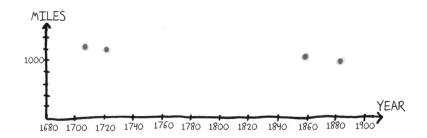

Okay, our data is a little sparse, but the descending pattern is clear. These days, statisticians have a favorite technique for embellishing such patterns: a tool known to economists, epidemiologists, and hasty generalizers everywhere as "linear regression."

First, we locate the "central point" of the graph. Its coordinates are the simple average of the existing data's coordinates.

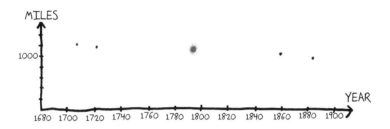

Then, from among all the lines traveling through this point, we pick the one that best fits the data, passing closest to the points already established.

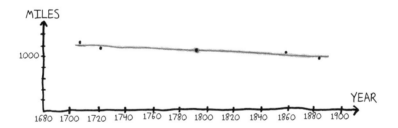

Voilà! We have now made the leap from a few scattered points—stubborn, stationary things—to a magnificent and continuous line. It comprises *infinite* points, and can be extended in either direction as far as we like.

For example, we can extend the line into the distant past:

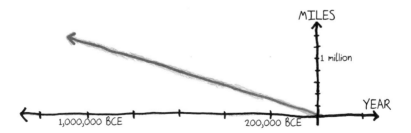

Lo and behold! A million years ago, the Mississippi was a cosmic monstrosity, more than a million miles long. Twain settled for a timid and inadequate image with his "fishing-rod" over the Gulf of Mexico. The *true* Mississippi jutted out five times farther than the moon, and whenever our stony satellite passed by, the Mississippi doused it like a fire hose.

Since lines go two directions, we can push our linear model forward in time, too:

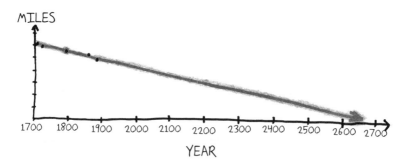

There you have it! Shortly before the dawn of the 28th century, the Mississippi will shrink to less than a single mile long. To accommodate, the North American continent will scrunch up like balled paper, allowing Cairo and New Orleans their long-awaited adjacency along the river. Between them will loom a crevasse 500 miles deep, plunging well into the Earth's mantle.

I hear you complaining. "No serious mathematics," you say, "could possibly be built upon such a wobbly foundation."

Ha! What is this "serious" mathematics? Math is logical play, the tomfoolery of abstraction. And in many of its games, straight lines are indispensable simplifiers. They help bypass slow meanders of calculation, like a cutoff that shortens a river. That's why straight lines crop up everywhere—in statistical models, in higher-dimensional transformations, in exotic geometric surfaces, and, most of all, in the essence of the derivative.

Take the parabola. If you possess eyes like a well-caffeinated eagle, and you give the illustration below a vigorous squint, then you will behold a deep and subtle truth: a parabola is not a line.

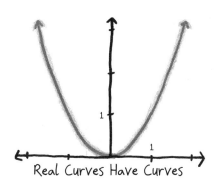

Real Curves Have Curves

Rather, it is—I apologize for lapsing into technical jargon—a curve. But let us zoom in closer. What do you see now?

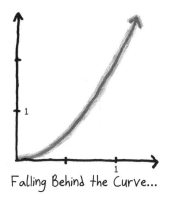

Falling Behind the Curve...

Still a curve, yes. But it is a curve less curvaceous, a parabola less parabolic. And observe what happens as we zoom in even closer:

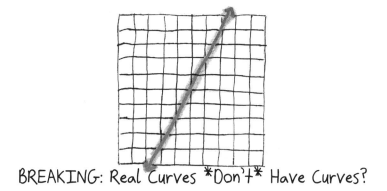

BREAKING: Real Curves *Don't* Have Curves?

The curvature is mellowing, gentling. We are lulling it to sleep. Zoom in close enough, and the curvature grows so slight that the naked eye can barely discern it. Technically, it remains a curve; for any practical purpose, it might as well be straight.

And at some infinitesimal scale—beneath all known sizes, though not yet at zero—the curve achieves that which we seek. It becomes, at least in our imagination, truly linear.

Now, what has all this to do with derivatives? Everything.

The derivative, you may recall, is a rate of change at a particular moment. For example, it might tell us how the length of the Mississippi River is evolving at this precise instant.

But the Mississippi's length does not change at a steady pace. It remains the same for a while, then abruptly shortens, then lengthens gradually. As human beings, we cannot begrudge the river for choosing a life of flow and flux, but as mathematicians, we certainly do. How can we abide such erratic riparian behavior? How can we speak of a rate of change, when the river won't adhere to any rate for more than an instant?

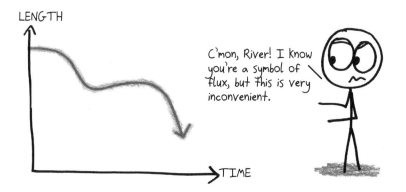

Simple: we can zoom in, as we did with the parabola. Go far enough, to infinitesimal scales, and the graph's curvature straightens out, allowing us to decipher the derivative.

Thus does all of differential calculus rest on one simple observation: *zooming linearizes.*

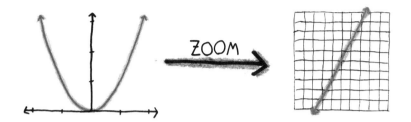

At large scales, the Earth is not flat. Indeed, our attempts to flatten it yield hopeless distortions like the Mercator projection, which makes Greenland (not even 1 million square miles) appear as large as Africa (almost 12 million square miles). But at small scales? Hey, why not! Zoom in far enough, and you'll never notice the curvature. If I'm tracing the Mississippi from Cairo, Illinois, to Columbus, Kentucky—a stretch of 20 miles, or just 0.08% of the way around the globe—then a flat map is perfectly adequate.

Twain commits the ancient sin of mistaking *local* linearity for *global* linearity. And what he did in jest, others have done in earnest. In his trenchant book *How Not to Be Wrong* (from which I pilfered many of this chapter's ideas), Jordan Ellenberg cites a remarkable specimen of wrongness: A 2008 paper in the journal *Obesity* asserted that, by 2048, the percentage of the US adults that are overweight or obese would have reached—drumroll, please—100%.

The researchers had extended their linear model too far, following it out into the vacuum of space, as the surface of reality curved away.

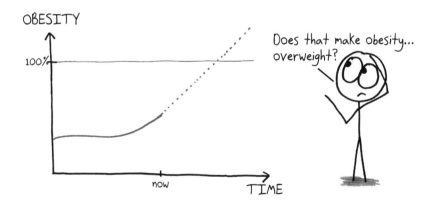

Another case study: In 2004, *Nature* published a brief paper noting that women's times in the Olympic 100-meter sprint had been improving faster than men's. Thus—"if current trends continue," the authors wrote with a wink—the women would overtake the men at the 2156 Olympics, when both sexes would run the race in a blistering eight seconds.

Alas, when the 2156 Olympics are held—in Space Paris, Moon York City, or the People's Republic of Google—I venture that "current trends" will no longer hold. That's because "current trends" always look linear, while history's full arc almost never does. Extrapolating the same model back to ancient Greece, we learn that the warriors ran the race in 40 seconds. That's a brisk walking pace, recently matched by a 101-year-old woman in Louisiana. The future looks even stranger, with a continual descent of gold medal times, culminating in *Star Trek* achievements:

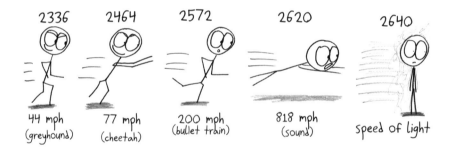

2336   2464   2572   2620   2640

44 mph (greyhound)   77 mph (cheetah)   200 mph (bullet train)   818 mph (sound)   speed of light

Life is like the Mississippi. It flows. It meanders. Zoom in close enough, and you may find a straight edge, but the whole landscape is a restless and ever-curving thing.

To close, I've got one last passage from *Life on the Mississippi*, about the sediment deposited at the river's delta:

> *The mud deposit gradually extends the land—but only gradually; it has extended it not quite a third of a mile in the two hundred years which have elapsed… The belief of the scientific people is, that the mouth used to be at Baton Rouge, where the hills cease, and that the two hundred miles of land between there and the Gulf was built by the river. This gives us the age of that piece of country, without any trouble at all—one hundred and twenty thousand years.*

Here is another linear model. Twain zooms in on the two most recent centuries, a geological instant during which the land has grown 1/3 of a mile—about 9 feet per year. Then, extrapolating backward, Twain concludes that 120,000 years ago, the delta lay 200 miles upstream.

Alas, Twain is making the same error as the *Obesity* researchers, the same error that he elsewhere satirized.

The Mississippi River as we know it dates to the end of the last ice age, a mere 10,000 years ago. Twain's linear model juts out another thousand millennia into the past, like a river poking out into the depths of space. He is asking a derivative to narrate all of eternity, forgetting that it speaks only for a single moment. ∎

MOMENT VI.

Sherlock Holmes struggles with kinematics.

VI.

SHERLOCK HOLMES AND THE BICYCLE OF MISDIRECTION

In "The Adventure of the Priory School" by Sir Arthur Conan Doyle, calamity strikes a posh English boarding school. A wealthy duke's 10-year-old son has vanished from the dormitory. Also missing: a German teacher, a single bicycle, and the commitment to serving a diverse population. With the local police befuddled, the school's desperate headmaster staggers into 221B Baker Street to enlist the help of fiction's most hallowed detective.

"Mr. Holmes," he says, "if ever you put forward your full powers, I implore you to do so now, for never in your life could you have a case which is more worthy of them."

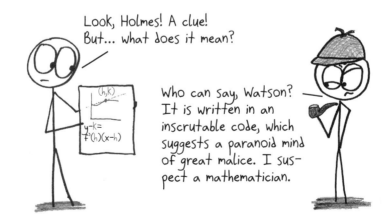

Look, Holmes! A clue! But... what does it mean?

(h,k)

$y-k= f'(h)(x-h)$

Who can say, Watson? It is written in an inscrutable code, which suggests a paranoid mind of great malice. I suspect a mathematician.

Hours later, Sherlock Holmes and Dr. Watson are prowling through a "great rolling moor" when they come across their first clue: "a small black ribbon of pathway. In the middle of it, clearly marked on the sodden soil, was the track of a bicycle." That's when Holmes launches into a classic piece of deductive reasoning:

> *"This track, as you perceive, was made by a rider who was going from the direction of the school."*

> *"Or towards it?"*

> *"No, no, my dear Watson. The more deeply sunk impression is, of course, the hind wheel, upon which the weight rests. You perceive several places where it has passed across and obliterated the more shallow mark of the front one. It was undoubtedly heading away from the school."*

What feats of physics! What genius of geometry! There is only one small problem, which the mellifluous prose disguises and which a simple diagram will make clear:

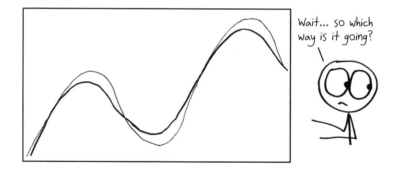

Here we see the thicker track crossing the thinner one. Is it thus apparent which direction the bicycle is traveling? No, alas, because Holmes has committed an uncharacteristic error. The rear wheel *always* crosses the front. This fact is no clue to direction, but a simple consequence of bicycle design, wherein the front wheel pivots while the back remains fixed.

How could Holmes leave us in the lurch like this? "Perhaps," suggests math professor Edward A. Bender, "he'd had some opium recently." One might blame Sir Arthur Conan Doyle, but I believe Holmes must assume responsibility for his own mistakes, like the fictional adult he is.

Lucky for the duke, there is a correct and elegant method for deducing a bicycle's direction from its tracks. It rests on a simple, potent concept from differential calculus: the *tangent line*.

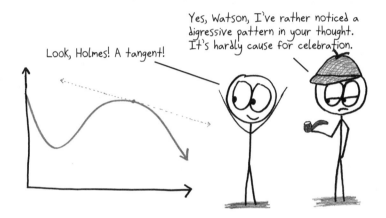

The word "tangent" grows from the same Latin root (*tangere*) as "tangible" and "tango"; all are words of touch, of caress. In mathematics, the tangent line grazes a curve at a single point. There, for a fleeting moment, it mimics the curve's instantaneous direction, its derivative.

For example, if the curve were the path of a car, then the tangent would indicate the direction of the headlights.

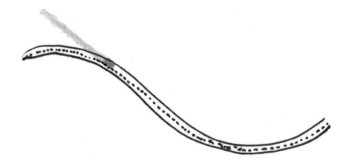

Or, for a more dramatic demonstration, tie a string around a stone, whip it in circles above your head, and wait for the string to snap. The stone will fly off in a straight line: the tangent to its path at the moment of its escape.

What now of bicycles? Because the back wheel is fixed in the frame, in any given moment it chases the front wheel. That is to say, its instantaneous direction of motion points toward the front wheel's present location.

Let us test-ride this fact with a preliminary puzzle: With no clues from track depth, can we still identify the front wheel?

Elementary, my dear Sherlock! Simply find a moment along one of the tracks where the tangent points off into space, gazing off in a direction that the bike never went. Would the back wheel let its attention wander so? Never! It keeps its eye trained on its partner at all moments. Thus, the track with outward-facing tangents must belong to the front wheel.

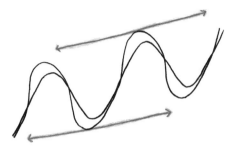

Now, the £6000 question, in honor of to the prize offered by the duke in the story: Which direction is the bicycle traveling?

There are only two possibilities to consider. First, suppose the bicycle moves from left to right. Draw the appropriate tangents for the back wheel, extending them until they intersect the front wheel's path:

The distance from back to front, along the tangent, ought to correspond to the length of the bicycle. But here, that distance fluctuates from point to point. We are left to conclude that, during its journey, the bicycle oscillated in length like a two-wheeled Slinky. Such a bicycle must belong to a rider of unparalleled skill and questionable judgment.

"The Adventure of the Priory School" offers a fitting comment:

> *"Holmes," I cried, "this is impossible."*
>
> *"Admirable!" he said. "A most illuminating remark. It is impossible as I state it, and therefore I must in some respect have stated it wrong.... Can you suggest any fallacy?"*

In our case, the fallacy is clear. We have left an alternative unexamined—that the bicycle is traveling from right to left:

Aha! These tangents, happily, are the same length. They reflect a bicycle of solid build and even more solid plausibility. We conclude that this was the bicycle's true direction.

Is this not a magnificent turn of reasoning? It extracts certainty from the residue of motion, plain truth from the coded language of geometry. It combines the close inspection of physical evidence with the careful exercise of abstract logic. In these qualities, it resembles every triumph of Holmesian inference—and, not coincidentally, of higher mathematics.

Holmes's relationship to mathematics is clear: it is his mirror image. That's why, when Sir Arthur Conan Doyle desired a nemesis worthy of the logical, sharp-eyed detective, he created a mathematician, Professor Moriarty. "The Napoleon of crime," he is described as "a genius, a philosopher," and "an abstract thinker," not to mention "the celebrated author of *The Dynamics of an Asteroid*, a book which ascends to such rarefied heights of pure mathematics that it is said there was no man in the scientific press capable of criticizing it."

It is discouraging to think what quick work Moriarty might have made of the bicycle tracks. Holmes's foe, we can trust, knows his tangent lines.

I myself first learned of the bicycle puzzle from Siobhan Roberts's lovely biography *Genius at Play: The Curious Mind of John Horton Conway*. In a memorable scene, three mathematicians team up to teach an experimental class at Princeton. It is a "sideways, subversive effort," targeting "math and poetry majors alike," and titled "Geometry and the Imagination." Expecting an enrollment of perhaps 20 students, they find themselves inundated with 92. As Roberts tells it, the kids got their money's worth:

> *The professors made up a ritual of entering the classroom en masse, sometimes with great pomp and circumstance, sometimes carrying a flag, sometimes wearing bicycle helmets, often pulling a red kiddy wagon heaped with polyhedra, mirrors, flashlights, and fresh produce from the grocery store...*

For one lesson, the professors "found large rolls of paper, tore off strips that were 6 feet by 20 feet at least," and then rode across them on bicycles with painted wheels. This created epic canvases of geometric bike art, life-sized puzzles. The students, like young Sherlocks, were tasked with determining which way the bicycles went.

But the professors had thrown in a wrinkle, one that might have puzzled even Moriarty:

> *A certain set of tracks, however, stumped the students. For that set, Peter Doyle [one of the professors] had pedaled up the sheet of paper and then back again, but on a unicycle.* ∎

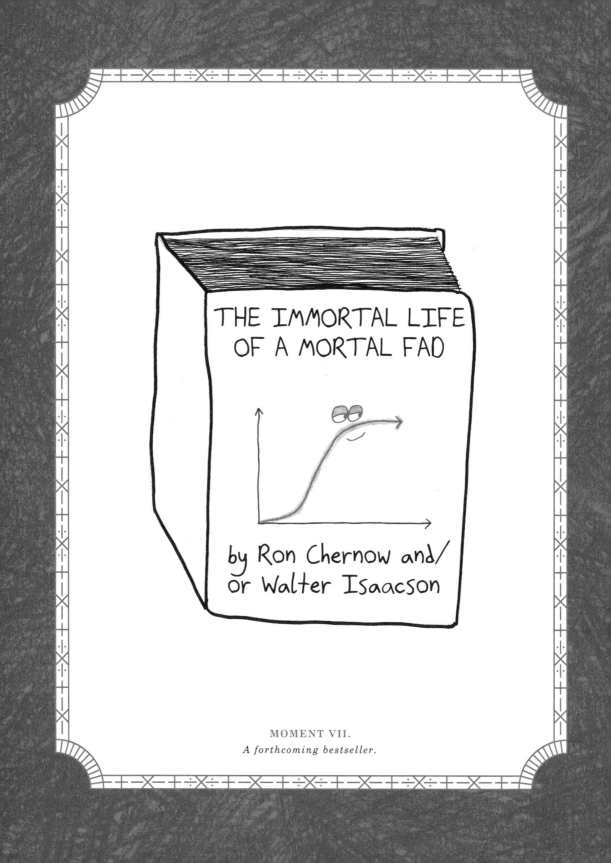

THE IMMORTAL LIFE
OF A MORTAL FAD

by Ron Chernow and/
or Walter Isaacson

MOMENT VII.
A forthcoming bestseller.

VII.

THE UNAUTHORIZED BIOGRAPHY OF A FAD

Thisis the story of a viral sensation. I leave it to you to decide which one: perhaps Hula-Hoops, or Rubik's Cubes, or Giga Pets, or the cheap Giga Pet rip-offs called iPhones. But it needn't be a toy! You can pick a linguistic change, or a technology, or a social network, or even a growing tumor or a population of rabbits. Whatever catches your fancy, and then, as fads do, catches everyone else's.

How? you ask through spittle and shock. *How can a chapter be so one-size-fits-all, so choose-your-own-adventurous?*

Well, because it's really the story of a curve. This curve:

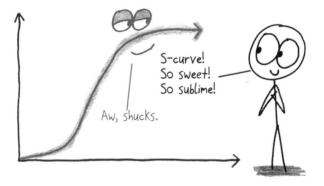

S-curve!
So sweet!
So sublime!

Aw, shucks.

This basic pattern, called *logistic growth*, is one of the great mathematical models around, not to mention a triumph of elementary calculus. And like all the classics, it unfolds in three acts.

First is Act I: Acceleration.

As we begin, our fad is not yet a fad, only a wild notion. *I will sell a rock as a pet,* some madman vows. Or perhaps: *I will choreograph a dance for the arms, and the whole world shall cry, "Hey, Macarena!"* Or even: *I will computerize a book of faces, for I am become Zuckerberg, Destroyer of Worlds.*

Visionary? Perhaps. But at the start, growth looks slow.

Things are not as bleak as they seem. During these inauspicious beginnings, the growth is actually—to a close approximation—*exponential*.

The word "exponential" has infiltrated the common language as few mathematical words have. ("Inner product" and "bipartite graph" still languish in tragic obscurity.) However, as always happens when an indie band makes it big, some of the original flavor and specificity has been lost. Folks toss around "exponential" as a synonym for "very fast," but its technical meaning is something more wonderful and precise: *when a thing's growth is proportional to its size.*

In other words: the bigger the object, the faster it grows.

In *linear* growth, you gain the same amount each time period. It may be slow, like a tree adding one ring per year. Or it may be fast, like a mutant Jack-and-the-beanstalk tree adding one ring per millisecond. What matters isn't speed, but consistency. If the growth rate never changes, then it's linear.

Contrast this with, say, a startup whose revenue grows 8% per month. At first, it's 8% of a trifling amount—a trifle of a trifle. But as time goes by and the company expands, that 8% growth refers to larger and larger numbers. In nine months, revenue doubles, and within a decade, the firm has metamorphosed from a $1000-a-month caterpillar into a behemoth $8-million-a-month butterfly. Give it another decade, and it's a $1-trillion-a-month monstrosity, accounting for 15% of global GDP. *That's* exponential.

You can capture the essence of the distinction in two brief equations:

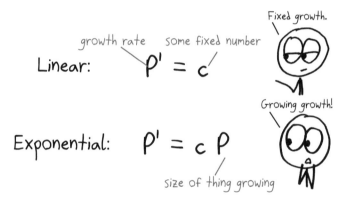

Now, this exponential honeymoon can't last forever, or every fad would consume the universe. In reality, this has happened only twice so far, with Beanie Babies and dabbing. Before long, we must enter Act II: Inflection.

Like "exponential," the phrase "inflection point" has oozed out of the mathematics textbook and into the general language. Myself, I always applaud the viruslike spread of math jargon, but I must point out that the popular usage—as "a moment when growth suddenly takes off"—gets inflection points rather backward.

In logistic growth, the inflection point is not when rapid growth *starts*. It's when rapid growth *climaxes*, hitting its fever-pitch maximum—and thus beginning a long, slow decay.

A derivative, you may recall, tells us how a graph is changing. Positive derivative? It's increasing. Negative? It's decreasing.

The *second* derivative tells us how the first derivative—i.e., the growth rate itself—is changing. Positive second derivative? Then our growth is speeding up. Negative second derivative? It's slowing down.

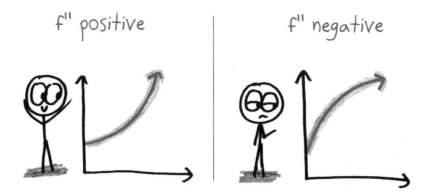

An inflection point is a moment of transition, when the second derivative changes signs: negative to positive, or (as in logistic growth) positive to negative. After building and building speed like a runaway train or an overplayed pop single, the acceleration finally stops and the growth begins—at long last—to slow.

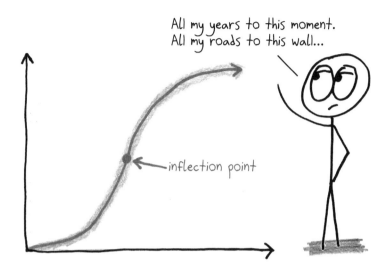

This is a special moment in the life of a fad: triumphant and bitter-sweet, as all peaks are. In the case of, say, Instagram, it's the month when the most users join. The network is not yet enjoying its *widest* adoption, but rather its *fastest* adoption, spreading quicker than it ever has before or ever will again. According to the graph, what comes afterward is a mir-ror image of its earlier trajectory: every moment of acceleration is now reversed with a corresponding deceleration.

This brings us to Act III: Saturation.

Here, the fad outgrows coolness. Parents know about it. Grandparents know about it. Even those pop-culture lepers called math teachers may scuttle crablike onto the bandwagon. Early adopters feel their pride turn to contempt. As prince of paradox Yogi Berra observed, "Nobody goes there anymore. It's too crowded now."

In the exponential pattern, growth is proportional to size. Logistic growth adds a crucial wrinkle: the growth is proportional to size, as well as to *the distance from some maximum size.*

The closer to the maximum, the slower the growth.

AHHHHHH, unchecked growth!

Exponential: $P' = cP$

growth rate   size of population

Ahh. Checked growth.

Logistic: $P' = cP(Max. - P)$

distance from maximum)

A forest can endure only so many rabbits; an economy, only so many electric cars; a human eye, only so many viewings of "Gangnam Style." Every system has finite resources, in one way or another. Facebook can never grow past the size of the human population, not unless it relaxes its prejudicial ban on dolphin and gorilla users.

For an illustration, we turn to chemistry, and the realm of *autocatalytic reactions*.

Chemistry studies reactions of all kinds, such as "explosive," "fizzy," and "ooh, pretty colors." Sometimes, there's a special molecule that speeds up a reaction, like a helpful assistant. We call these "catalysts."

A few very special reactions produce their own catalysts. This creates a positive feedback loop: the more catalysts you generate, the faster you generate more catalysts. Faster and faster, the reaction fizzes and/or explodes...but such cycles cannot last forever. As the original stockpile of inputs dwindles toward zero, we are left with loads of catalysts, but nothing left to catalyze. The reaction slows.

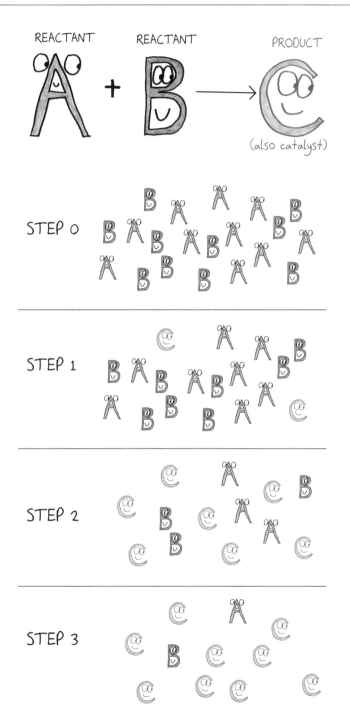

Fads follow similar logic. The more people a fad recruits, the more people *they* recruit. A feedback loop leads to exponential growth—at least, for a little while. But sooner or later, the fad begins to exhaust its targets. Lots of recruiters, and no one left to recruit. A surplus of catalysts, and nothing left to catalyze.

You might say, if you're a fan of ostentatious chemistry jargon, that a viral fad is an autocatalytic reaction of humans.

Mathematical models fall into two camps. *Mechanistic models* embody the principles of the original, like a model plane with a scale engine. Meanwhile, *phenomenological models* aim only for surface-level resemblance, like a model plane that looks cool and genuine but cannot by any means fly.

Before test-piloting the model in this chapter, you might fairly ask: Which kind is it?

Silicon Valley would tend to say "mechanistic." Specify a "virality coefficient" (for the number of new people each fad follower brings into the fold), estimate the size of the whole market, throw together a PowerPoint deck, and, boom, you're ready to pitch investors.

On the other hand, consider the true story of the eminent biologist who decided to forecast, via logistic growth model, the future of the US population. He sized up early 20th-century data, cracked his knuckles, and concluded that the nation would stabilize at just under 200 million people, a figure we passed, oh, about 120 million people ago.

If it can't help us predict where a fad will level out, then what good is the logistic model? Is it a mere folktale, a just-so story in graph form?

Perhaps. But don't underrate that. You and I are storytelling creatures. Narratives structure our actions, our thoughts, our takeout orders. Even if it cannot predict, the mythologized tale of logistic growth can enrich our thinking, highlight key moments, and hint at possible futures.

Mathematical models gesture at a reality too complicated to express in full. A little simplification is a healthy human response—as long as we read the fine print before playing with the toy. ■

MOMENT VIII.

A puzzle refuses to yield.

VIII.
WHAT THE WIND
LEAVES BEHIND

I t's a bright November day in Massachusetts. The wind is sweeping
leaves from the trees, like winter taking down the autumn decora-
tions. I'm cradling a cup of tea and describing this book—currently
just a slapdash outline and a few half-baked paragraphs—to an English teacher friend, Brianna. It's a tour of calculus, I explain, but with no
fancy equations. No intricate computations. Just the ideas, the concepts—
all illustrated by stories. The stories will cut across human experience,
from science to poetry, from philosophy to fantasy, from high art to every-
day life. It's easy to rhapsodize when I haven't written it yet.

Brianna listens well. She is, by her own description, "not a math person," and is, by my description, curious, thoughtful, incisive. Pretty much
the exact reader I'm hoping to reach. As we chat, something occurs to her,
a riddle that a math-teacher colleague once rolled out. She grabs a piece of
paper and draws a rectangle.

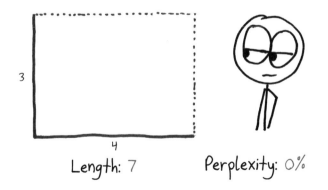

Length: 7   Perplexity: 0%

"How long is the dotted part?" she asks.
"7 inches," I say. "3 plus 4."

"Okay," she says, "now what about this?"

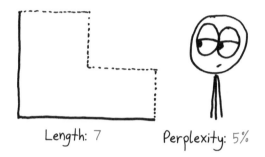

Length: 7   Perplexity: 5%

"Still 7," I reply. "The two horizontal bits add up to 4; the two vertical bits add up to 3. Breaking them into multiple pieces doesn't change the overall length."

"Right," she says. "So how long is the dotted part *now*?"

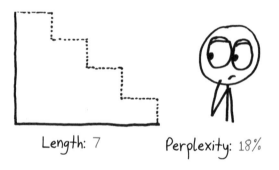

Length: 7   Perplexity: 18%

"Still 7," I say, "by the same logic."
She draws again. "And now?"

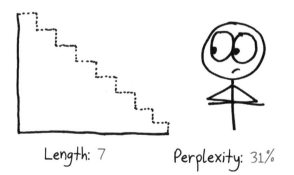

Length: 7   Perplexity: 31%

"7…"

"Okay, so what if we make *infinite* steps on the staircase, and create a shape like this?" she says.

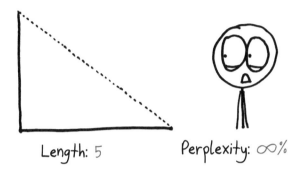

Length: 5   Perplexity: ∞%

I frown. It's the Pythagorean theorem, the oldest rule in just about any book: $a^2 + b^2 = c^2$. In this case, with $a = 3$ and $b = 4$, we're left with $c = 5$. I say so.

"5. Yes, exactly." Brianna drops the pencil like a mic. "So what's going on here?"

In the living room with Brianna are the following humans: (1) her husband, Tyler, a former calculus teacher and current data-science entrepreneur; (2) my wife, Taryn, a research mathematician; and (3) me, a guy who writes books about math. Between us we have more than 40 years of math education, with degrees from MIT, UC Berkeley, and Yale. We know all about limits and convergence, the geometry of approximation. We know 7 is not 5.

But in the face of this puzzle, we freeze. I feel like the cosmos is trolling me—reaching around my back to tap me on the opposite shoulder, so that I look the wrong direction. I can almost hear it chuckling. Or maybe that's just the wind.

"Nonuniform convergence?" is Taryn's cryptic mutter.

"It's not a valid limit," Tyler says with unconvincing confidence.

In my own thoughts, several possible refutations elbow for space, none of them the least bit illuminating or explanatory.

All I can say is, "Huh."

Brianna's puzzle strikes right at the heart of calculus, at the underpinning philosophical concept called a *limit*. A limit is the final destination of an infinite process. You don't necessarily reach a limit; you approach it, closer and closer—closer than words can write or imagination can draw. Brianna is taking a limit here, a real sneaky one. It points, in some paradoxical way, toward two destinations at once. Every step of the way, the length is 7; then somehow, right at the end of eternity, it's 5.

Paradoxes like this have long plagued calculus. A generation after Gottfried Leibniz and Isaac Newton first developed it, the philosopher George Berkeley scolded them for sloppy thinking. Newton had claimed to inspect values not *before* they vanished (when they were still finite numbers), nor *after* they vanished (by which point they'd be zero), but *as* they vanished. What does that even mean?

"And what are these…evanescent increments?" Berkeley jeered. "They are neither finite quantities nor quantities infinitely small, nor yet nothing. May we not call them the ghosts of departed quantities?"

Brianna's is hardly the only such paradox. Another version begins with an equilateral triangle. Assuming all three sides are equal, the red path is twice as long as the black path.

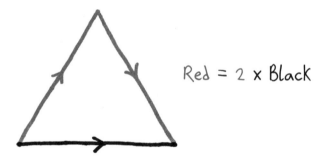

Next, take the two red edges, snap them each in half, and thereby turn our up-and-down path into an up-and-down-and-up-and-down path.

The red length hasn't changed; we've only rearranged its pieces. It thus remains double the black. And we can repeat this process—snapping and rearranging, snapping and rearranging—with the red length remaining double the black one every step of the way.

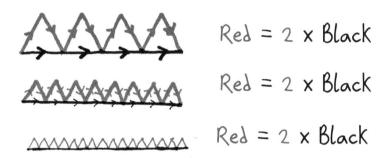

If we repeat the snapping infinite times, then the original red tent crumbles into a straight line of dust, indistinguishable from the black. But...doesn't that make the path double its own length?

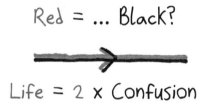

It took centuries of stumbles and pratfalls for researchers to map this terrain. "Reading mathematicians from that period," writes professor William Dunham, "is a bit like listening to Chopin performed on a piano with a few keys out of tune: one can readily appreciate the genius of the music, yet now and then something does not quite ring true."

The baffling truth—no less baffling for its no-duh simplicity—is that not everything survives the limit process.

Take the sequence 0.9, 0.99, 0.999, 0.9999... Every step of the way, you have a fraction, a *nonwhole* number. And yet somewhere down the yellow brick road of infinity, the sequence converges to 1.

Does that mean 1 is not whole? Egad, no! It simply means that the destination need not resemble the path that brings you there. Wooden stairs may lead to a carpeted landing.

Here's an example my wife teaches in her intro-to-analysis classes: a triangular wave moving across the flat, still water of the x-axis.

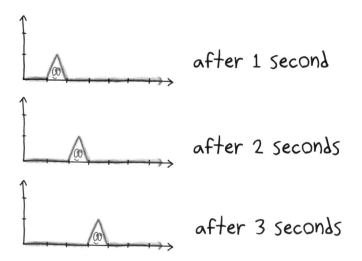

Each point is zero for a while, then briefly nonzero as the wave passes through, then zero again, forever and ever more. Each point, therefore, converges sooner or later to a height of zero. That means the limit of the whole scene is a horizontal line, the x-axis.

But what happened to the wave? Does the limit wipe it from existence, like a neutron bomb?

In a word: yes. Limits can do that.

You never really "reach" a limit. Approach it, sure—so close you can smell it, so close that it aches—but you don't arrive. The jump to the limit is an act of transcendence; it's akin to the leap of mortality, the transformation from time-bound body to timeless soul. Why should every property survive the journey? Our bodies have hair and teeth; do we expect an afterlife of hairy, toothy spirits?

The miracle of calculus, the unfathomable secret of the whole discipline, is that so much *does* survive the mortal leap. The derivative and the integral are both defined by limits. Yet they don't collapse in paradox. They work.

Puzzles like Brianna's drove math forward in the 19th century. A whole generation of scholars worked in concert to eliminate paradox from calculus once and for all. This meant transforming the intuitive, geometric work of their predecessors into something ironclad and rigorous, a reconceptualized calculus that preserved some features of the original while forsaking others.

That's how it is with limit processes. Some facts vanish like autumn leaves; others endure like winter branches. ∎

MOMENT IX.

A particle struts its stuff.

IX.

DO THE DUSTY DANCE

The year: 1827. Our protagonist: a jolly gray-haired botanist named Robert Brown. He hunches over a microscope, gazing at a slide of wildflower pollen, which, decades before Netflix, qualifies as weekend fun. Here, on his pollen-dusted microscope slide, Brown notices something rather peculiar:

A miniature dance party.

Tiny particles, emitted by the pollen grains, are oscillating before his eyes. They jitter. They jive. They hop around like kernels of popcorn, or caffeinated rabbits, or me at a friend's wedding. They wiggle as if privy to a secret broadcast of "Uptown Funk." What fuels this frantic activity?

♫ DON'T BELIEVE ME, JUST WATCH ♪

Is it perhaps the lively force of the pollen, the sperm-like wiggling of floral sex cells? Nope. For one thing, even when the liquid is sealed in glass for an entire year, the dancing never stops. (The true test of a dance party.) For another, Brown finds the same motion in particles of glass, granite, smoke, and even in dust drawn from the Great Sphinx of Giza, which suggests that historic sites were chiller back then about tourists waltzing off with free samples.

Brown isn't the first to gawk at this phenomenon. A generation earlier, a scientist named Jan Ingenhousz noticed coal dust quivering on alcohol. Nearly two millennia prior to that, the Roman poet Lucretius had written of dust specks catching in the light. This choreography is ubiquitous and ancient.

So what exactly is it?

Well, the world is made of atoms. These are in constant, jostling motion. Without an electron microscope, we can't see atoms, but we *can* see larger particles bombarded by them—particles like sphinx dust and wildflower pollen. Picture the giant ball at Disney's Epcot, under a ceaseless barrage from trillions of invisible marbles, and you'll get the idea.

In any given moment, by random chance, the bombardment from one side slightly outweighs the bombardment from the opposite side. That causes the particle to leap in one direction. The next moment, the pattern shifts, causing the particle to bound off in a new direction.

This carries on into eternity, moment by moment by moment by moment.

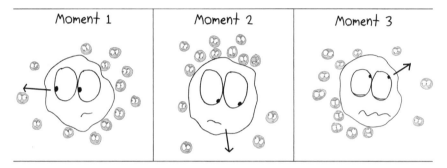

and so on, forever...

The jig of the particles—dubbed "Brownian motion"—exhibits puzzling features. It is *indiscriminate*, with particles favoring no direction over any other; *independent*, with each particle dancing alone, showing no correlation with its neighbors; and *unpredictable*, with past movements offering no hint of future ones. But perhaps strangest of all is the nature of those direction changes.

They are, in our mathematical model, *nondifferentiable*.

That term requires some explanation, so suppose you are a baseball. Suppose that I fling you into the air at a speed of 25 meters per second. Suppose you forgive me for this aggression, and together we wonder: What happens now? Will you puncture the atmosphere, thence to drift lonely and errant among the stars?

Fear not, my red-stitched friend! You are a citizen of Earth, subject to the planet's gravitational pull. Thus, after a second, you have slowed to 15 meters per second. A second later, you are down to 5 meters per second. Over the next half second, you slow even further, until at last you reverse direction, and begin to accelerate downward.

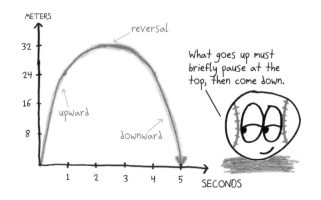

At the apex of this journey comes a peculiar and wondrous moment, when you have ceased to rise but have not yet begun to fall. For this brief hiccup in time, you are motionless, "traveling" at a speed of zero meters per second.

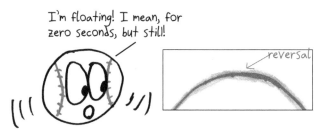

Now, what if we equip you with rocket boosters? Once a mere sphere of cowhide, you now become a jet-powered sphere of cowhide. You blast upward, then blast downward. Is this a different *kind* of direction reversal?

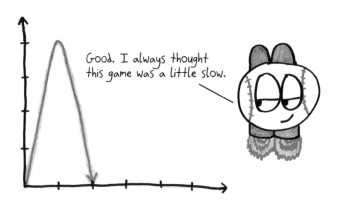

Not really. Sure, what took a whole second beforehand now unfolds in a fraction of one, but the basic pattern remains. After your upward motion has slowed and before your downward motion has begun, there comes a singular moment of reversal, when your instantaneous velocity is zero.

Hmm. Still not a very sharp turn.

Only by flexing our mathematical imagination can we imagine an alternative, like this one:

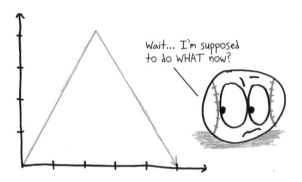

Wait... I'm supposed to do WHAT now?

Something bizarre is unfolding here. You move directly from "traveling up" to "traveling down," with no transition to speak of: no pause, no passing go, no seventh-inning stretch.

Even zooming in—our go-to maneuver for all things calculus—lends no clarity. No matter how close you look, or how much you slow down the video, that moment of reversal remains a very odd duck. A trillionth of a second prior, the baseball is traveling 10 meters per second upward; a trillionth of a second later, it's traveling 10 meters per second downward. There is no deceleration, no acceleration—just a sudden about-face, so abrupt and mysterious that the mind can scarcely grasp it.

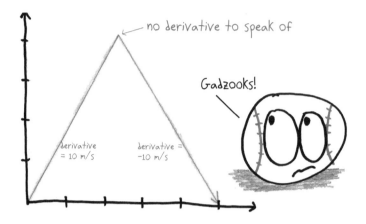

What speed is the baseball traveling at that moment? Actually, the motion is so pathological that the concept of velocity loses meaning. The baseball, for that instant, *has* no speed. In the jargon of calculus, its position function is *nondifferentiable*.

Now, with a sudden ricochet, let us bound back toward Brownian motion. What the baseball can never do, particles in Brownian motion appear to be doing daily. Perpetually.

An isolated point of nondifferentiability, a single sharp reversal in an otherwise smooth stretch, is bad enough. But half a century after Brown, analyst Karl Weierstrass constructed a mathematical function far more frightening. He didn't settle for just one nondifferentiable point, or two, or 20. He built a function that was nondifferentiable *everywhere*.

On the graph of Weierstrass's Frankenfunction, every single point is a sharp corner.

Struggling to visualize it? You and me both, pal. The best I can offer is some approximations: the first few steps on an evolutionary ladder that leads, via infinite ascension, to Weierstrass's demonic porcupine.

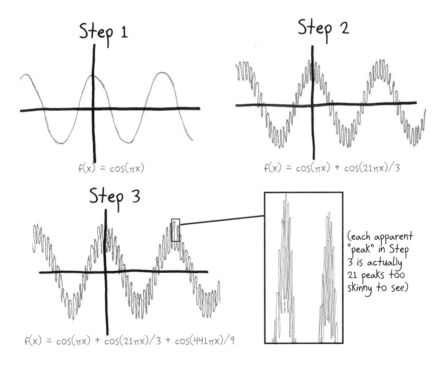

Let us be clear on the nature of this horror. It is a single, unbroken curve, with no jumps or gaps. But it is so janky and jagged that neither human hand nor graphing software can sketch it. This unimaginable monster, writes mathematician William Dunham, "drove the last nail into the coffin of geometric intuition as a trustworthy foundation for the calculus."

Émile Picard, a French mathematician, lamented the change: "If Newton and Leibniz had thought that continuous functions do not necessarily have a derivative," he warned, "the differential calculus would never have been invented." Another French mathematician, Charles Hermite, put it even more grimly: "I turn away with fright and horror from this lamentable evil of functions that do not have derivatives."

In the history of calculus, Weierstrass's pointy-faced devil marks a sharp turning point, an abrupt and definitive change of direction.

The Historical Path of Calculus

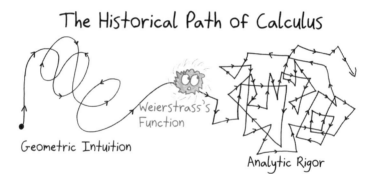

Geometric Intuition

Weierstrass's Function

Analytic Rigor

At times like this, it can seem that math has lost its head in the clouds, and then lost the clouds up its own butt. Who cares about these impossible technicalities, these unimaginable abstractions? Is Weierstrass just chasing philosophical claptrap for its own sake, disregarding math's prime directive to be—y'know—useful?

Guilty as charged. "It is true," said Weierstrass, "that a mathematician who is not somewhat of a poet, will never be a perfect mathematician."

And yet, if you've grown accustomed to sharp reversals, perhaps you see where this is going. This nowhere-differentiability, the feature that makes Weierstrass's spiky pet so horrifying and unreal, the trait that freaked out a whole generation of mathematicians...well, that's exactly how our model of Brownian motion works, too.

The path of a particle in Brownian motion doesn't exhibit just a few sharp corners. Its whole life is sharp corners. Every instant, a new and totally unpredictable step unfolds in the frenetic dance of the universal dust. Since a derivative is just a speed, Brownian motion is a kind of speedless motion, a buzz of activity that conventional calculus cannot describe, except to say "Wow" and "Whoa" and "*What?*"

I love the strangeness of Brownian motion: the path no hand can draw, the motion no speed can name. Is this why the authorities let Brown abscond with a stone from the Great Sphinx? Perhaps they sensed that his work, like the sphinx's, and like calculus, was a game of paradox and ancient riddles, a thing both impossible and entirely real. ■

MOMENT X.
The dangers of data visualization.

X.

THE GREEN-HAIRED GIRL AND THE SUPER-DIMENSIONAL WHORL

S ometime in the future, when folks vacation on Mars and sport green hair flecked with pearl dust, there is a happily married wife named Oona. Her husband, Jick, is "the sweetest man in the solar system," although the primary evidence for that claim is that "he always remembered anniversaries," so perhaps it isn't the most competitive solar system for men. To his further credit/debit, Jick strove to "share his interests" by giving Oona regular lessons in mathematics.

> He'd spent the quieter moments of their honeymoon (they'd taken an inexpensive stratosphere trip around the world) trying to teach her about calculus.

> He explained everything, he explained all about everything right from the start clear through to the end. He explained so much she got mixed up listening to him.

As a fictional character in a 1948 short story, Oona didn't have access to the term "mansplain." Instead, she accepted Jick's lessons and counted herself lucky: "Gee," she thought, "lots of husbands never spoke to their wives except to complain about the cooking."

One day, Jick comes home with a triumphant gift: "the finest robot brain ever yet devised," an appliance called "the Vizi-math." He explains:

> "You write any mathematical expression on a piece of paper, feed it into the machine... What you get is a translation into visual terms of the expression you were interested in."

Unlike Jick's other gifts to Oona, the Vizi-math actually helps. And for a high-tech dream machine, it helps in a rather simple way: by showing how multiplication is all about rectangles.

For example, the fact that $5 \times 4 = 20$ is best understood by considering a 5-by-4 rectangle:

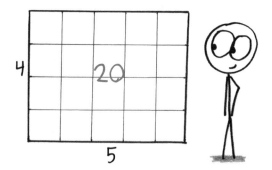

This works for multiplying nonwhole numbers, too, such as 6×2.5. You get 12 whole squares, plus six half squares, for a total area of 15.

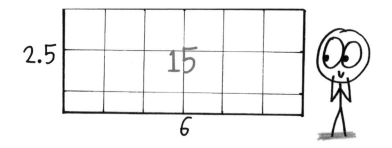

It even explains how the operation of "squaring" earned its name: because multiplying a number by itself creates a square.

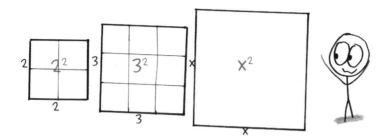

"To learn mathematics without pictures is criminal," said Benoit Mandelbrot, "a ridiculous enterprise." But somehow, teachers like me often fail to follow through. In this futuristic 21st-century world we inhabit, where tools like Wolfram | Alpha and Desmos could put Vizi-math to shame, we remain an all too Jick-like profession.

Take one of the first rules taught in any calculus course: that the derivative of x^2 is $2x$. I must admit that when teaching this fact, I always trot my students along the algebraic path:

Little marks! Why do you torment me so!

$$\lim_{\Delta x \to 0} \frac{(x+\Delta x)^2 - x^2}{\Delta x} = \lim_{\Delta x \to 0} \frac{x^2 + 2x\Delta x + (\Delta x)^2 - x^2}{\Delta x}$$

$$= \lim_{\Delta x \to 0} \frac{2x\Delta x + (\Delta x)^2}{\Delta x}$$

$$= \lim_{\Delta x \to 0} 2x + \Delta x$$

$$= 2x$$

Why am I such a Jick about this? Is it that standardized education leads inexorably to the rote? Is it that we teachers don't *know* good visuals, being products of the system ourselves? Or is it the lingering influence of Bourbaki, a radical 20th-century collective of mathematicians whose battle cry "Death to triangles!" warned that visual intuition is misleading and abstract symbolism is the only ground to stand on?

Whatever the cause, the Vizi-math offers an alternative. Begin with a square, x by x:

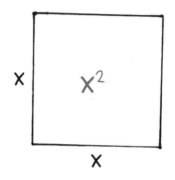

The "derivative," as you may recall, is an instantaneous rate of change. The operative question is: "If we change x by a little tiny bit, how much would x^2 change by?"

So let's go ahead and enlarge x by a tiny bit, called dx:

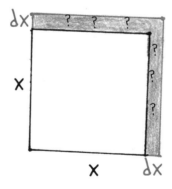

The growth in x^2 consists of three regions: two long, skinny rectangles (each x by dx) plus a tiny little square in the corner (dx by dx).

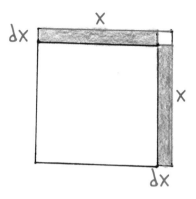

We pause here to reflect on the nature of tininess. Like, there's tiny, and then there's *tiny*. Say x is 1 and dx is $\frac{1}{100}$; pretty small, right? Sure, but $(dx)^2$ is a hundred times smaller: just $\frac{1}{10,000}$. It's a tiny so tiny the old tiny looks huge.

And what if dx is even smaller, like $\frac{1}{1,000,000}$? Then $(dx)^2$ is a million times smaller, equaling a mere $\frac{1}{1,000,000,000,000}$. That's a whole new tier of tininess.

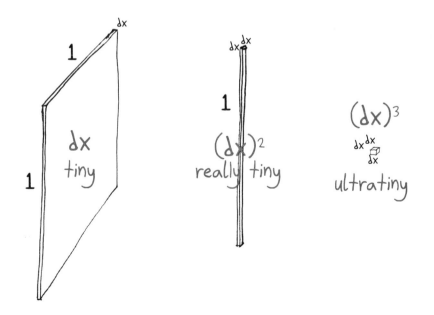

What is dx really? Well, it's infinitesimal, smaller than any known number. (John Wallis, inventor of the infinity symbol, wrote infinitesimals as $\frac{1}{\infty}$, though your teacher may bristle at this notation.) So $(dx)^2$ is not just a hundred or a million times smaller; it is *infinitely* smaller, an infinitesimal of an infinitesimal. It might as well be zero.

So how much has x^2 grown? Ignoring the negligible $(dx)^2$, it has grown by two rectangles—one for the width, and one for the height.

Hence, the derivative is *2x*.

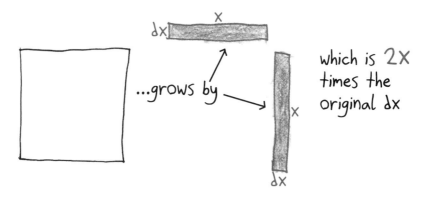

Back in the living room of the future, Oona finds this demonstration enthralling.

> So that *was what they meant when they talked about squaring a number. Not just multiplying it by itself, or whatever Jick had said, but turning it into an honest-to-goodness square… Mathematics, then, wasn't just a lot of numbers and letters and foolishness. It meant something, and a mathematical expression was like a sentence with something to say.*

Energized, Oona dives into the Vizi-math's next demo. As any calculus student can attest—with either a yawn or a sob—the derivative of x^3 is $3x^2$. Oona's question—and mine, and my students', and, heck, the question of any person ever subjected to the tortuous algebra of a well-intentioned Jick—is *Why?*

Vizi-math knows. Just as x^2 gave us a square, x^3 yields a cube:

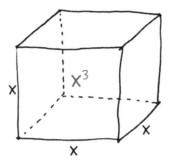

Again, allow x to grow by a little increment dx, and observe: every side of the cube grows, too.

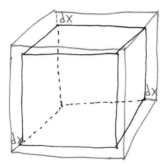

This creates lots of new regions. First, there are three flat squares, their depth infinitesimal:

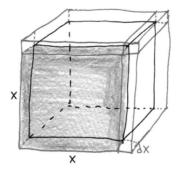

Next, there are three skinny rods, their depth and width *both* infinitesimal:

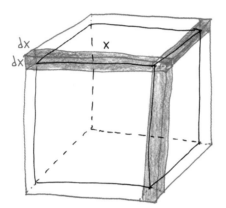

And in the corner sits one supertiny cube, its depth, width, and height *all* infinitesimal:

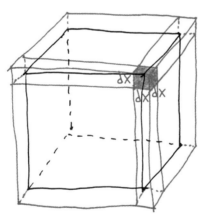

These latter four objects—the twiglike rods and the cutesy cube—are smaller than small, tinier than tiny. They amount to practically nothing when compared to the sheetlike square regions. Thus, the derivative: when you grow the edge of a cube, the cube grows by three flat squares, one per dimension.

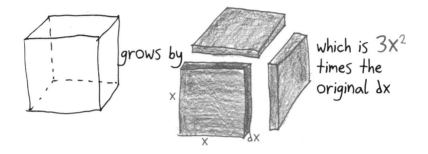

Watching this unfold on the Vizi-math, Oona eats it up:

> *Now that she knew she could understand mathematical things after all, there was a sort of bubbling elation in her veins.*
>
> *No more seeing Jick look so disappointed and hurt, no more listening to him talk, talk, talk, while she wondered what in the System he was talking about. From now on she'd take her troubles to the Vizi-math.*

All this unfolds in the story "Aleph Sub One," by Margaret St. Clair, a forgotten peer of Asimov, Bradbury, and Clarke. Her work mixes techno-optimism with social pessimism. Oona's is a timeline where the appliances get better and better yet the people stay forever the same. "I like to write about ordinary people of the future," St. Clair once explained, "surrounded by gadgetry of super-science, but who, I feel sure, know no more about how the machinery works than a present day motorist knows the laws of thermodynamics."

In this regard, "Aleph Sub One" stands out. The Vizi-math, unlike a cleaning or cooking device, endows Oona with something that really matters: understanding. She can now look past opaque formulas to the meaning beneath. She can see the light at the end of Jick's long, dark lectures.

It's quite a vision: feminist liberation through geometric visualization.

"The battle between geometry and algebra is like the battle between the sexes," mathematician Sir Michael Atiyah once said. "It's perpetual...

The dichotomy between algebra, the way you do things with formal manipulations, and geometry, the way you think conceptually, are the two main strands in mathematics. The question is what is the right balance."

Oona may inquire: Balance? Why do we need the muddle of algebraic symbols?

Because geometry has limits. The derivatives of x^2 and x^3 are easy enough to picture, but for x^4, you'd need to sketch a four-dimensional tesseract. Good luck. Oona tries it on the Vizi-math, to little effect: She sees "a thing like a cube, like a bunch of cubes, a thing that made her eyes smart." It vanishes a moment later, having surfaced "so briefly that Oona couldn't be sure whether she had really seen it or not."

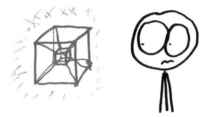

That's when Oona makes a fateful decision to feed the Vizi-math the gnarliest, most complicated expression she can devise.

She wrote steadily for nearly five minutes, sprinkling her work liberally with dx*'s,* n*th powers, and a good many* e*'s... Under her equation, in her round, childish penmanship, she wrote* N = *five.*

When the machine sputters and goes blank, Oona shrugs, and leaves to run some errands. When she returns, it's to a home engulfed in an "unnatural reddish blur, a thing that rotated slowly and was shaped like the whirlpool you get when water runs out of the sink." The attempt to display her monstrous equation has created a space-annihilating vortex.

It rings true to me: nonsense math symbols seem quite capable of destroying reality.

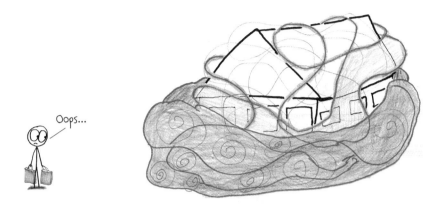

In the end, Oona manages to save the day by slipping a handwritten note into the whorl: "I made a mistake. I'm sorry. *N* doesn't equal *five*. Zero (0) is what *n* is equal to." The Vortex receives her note, and "for a moment the universe seemed to wobble on the edge of an abyss. Then it appeared to shrug its shoulders and decide to settle down."

Perhaps not all derivatives are meant to be visualized. ∎

MOMENT XI.
A small thing with big possibilities.

XI.

PRINCESS ON THE EDGE OF TOWN

Once upon a time, perhaps 29 or 30 centuries ago, there lived a princess named Elissa. According to surviving texts, her brother Pygmalion was "quite a boy." This is the polite way of saying that he murdered Elissa's husband for gold.

With wits, guile, and no doubt some newly developed trust issues, Elissa escaped across the Mediterranean to the African coast. She arrived with many followers but little to trade. Ever the hustler, she bargained for "a piece of ground, as much as could be covered with an oxhide."

It doesn't sound like much. But Elissa was a wily lady. As one source writes, "she directed the hide to be cut into the thinnest possible strips," and then, with an admirably flexible approach to contract law, she reinterpreted "could be covered with" as "could be enclosed by."

The stage was thus set for the most famous maximization problem of antiquity. How much ground can you fence off with some long strips of hide?

The puzzle is known today as an *isoperimetric problem*: the prefix *iso-*, meaning "same," and *perimeter*, meaning "sneaky lady." By etymological coincidence, *perimeter* also means "the length around the outside of a region."

The question is, of all the shapes you can enclose, which has the greatest area?

I don't know what units Elissa used. Probably not meters, unless she was a *very* early adopter, so let's say her oxhide strips amounted to 60 "ox-feet" (each defined as "1/60th of the total length that Elissa had").

Now, the cosmos of geometry houses limitless options for Elissa's city:

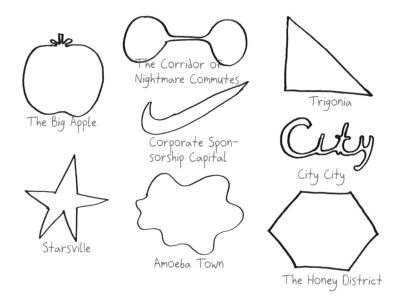

To avoid drowning in variety (or in gnarly area computations), let's consider the simplest class of shape: rectangles.

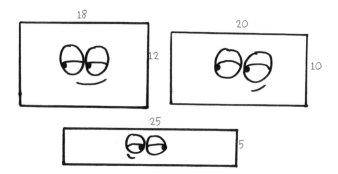

With limited oxhide, Elissa faces a trade-off. The more she extends the base, the more she must shorten the height, and vice versa. Bump one from 17 to 18; the other will drop from 13 to 12.

We can restate the area: instead of "Base × Height," call it "Base × (30 − Base)."

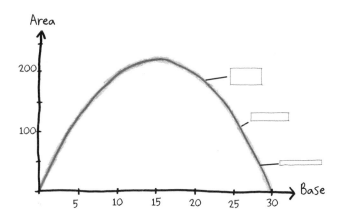

On the graph above, each point signifies a possible rectangle, a nascent empire for Elissa. At the far left, we find foolish plans, such as 1 × 29; at the far right are their mirror images, such as 29 × 1. Each of these proposals yields a paltry area of 29 square units, cramped enough to make Boston look spacious.

Why such feeble results? Just consider the derivatives. $\frac{d\,\text{Area}}{d\,\text{Base}}$ tells us the area's response to a change in base.

Meanwhile, $\frac{d\,\text{Area}}{d\,\text{Height}}$ tells us the area's response to a change in height.

Extend the base, and the area barely budges. Extend the height, and it soars. In differential terms, $\frac{d\,\text{Area}}{d\,\text{Base}}$ is puny, while $\frac{d\,\text{Area}}{d\,\text{Height}}$ is tremendous. That's the flaw of any such spaghetti-shaped rectangle, with its elongated base and stunted height. By design, it spends almost all of the precious oxhide on the stingy derivative, while allocating almost none for the generous one.

A smarter plan? Spend until the derivatives are equal. This happens—as the graph indicates—when the sides themselves are equal, in the 15-by-15 square.

OPTIMUM CHILL – noun. The state of calm that arises when all your derivatives are equal.

Have we solved Elissa's problem? Is it time to cut the red ribbon and begin haggling over parking spaces? Not so fast; the princess has another trick up her sleeve. Instead of spreading her oxhide in an open plain, what if she fences off a plot against the Mediterranean shore? That way, rather than share her oxhide among four sides, she need only allocate for three.

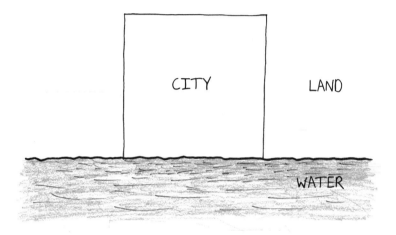

Previously, Elissa could afford a 15-by-15 square. Now, she can rope off a 20-by-20 region. Her area thus leaps from 225 to 400, a city birthing overnight suburbs. Surely, now we can elect a mayor and begin, at long last, to complain about construction?

Well, just to check, bring back the derivatives. Here's what $\frac{d\,\text{Area}}{d\,\text{Height}}$ tells us: an extra smidge of height brings an extra 10 smidges of area.

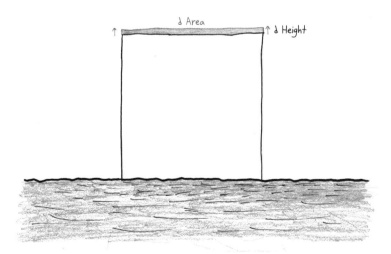

Not too shabby. And presumably it's the same with $\frac{d\,\text{Area}}{d\,\text{Base}}$?

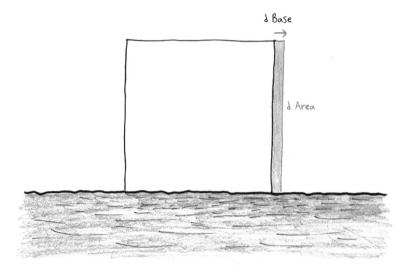

Egad! An extra smidge of base yields an extra *20* smidges of area! The derivatives are not equal!

On inspection, it makes sense. Here, every smidge of height entails two walls, whereas every smidge of base entails only one. Base is therefore "cheaper" by a factor of two. Roping off a square misallocates the resources; we seek instead a shape where the two derivatives are equal.·

Time to bust out another graph:

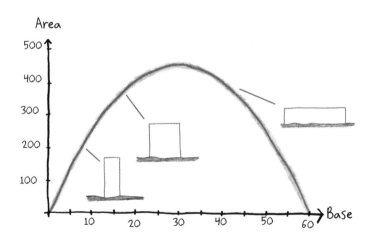

The maximum turns out to be a 15 × 30 rectangle, with an area of 450 square units.

A triumph by any measure. Elissa has converted her congested oxhide Manhattan into a sprawling oxhide Houston. And yet—and yet!—by drawing upon something called *calculus of variations*, which considers whole families of curves, Elissa can squeeze yet a little more area out of her unwitting negotiating partners. This brings us to the truly optimal solution: a semicircle whose diameter hugs the coastline.

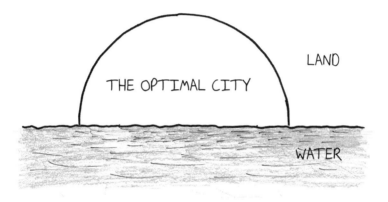

Its approximate area: 573 square units. Not bad for a day's optimization.

All this, the Roman historians tell us, took place in the late 9th century BCE. In the years to follow, that semicircular plot of land would grow into a prosperous and powerful port city called Carthage. It would reign as a superpower until Rome challenged its supremacy in three brutal wars. For years, Cato the Elder would end every one of his speeches with the words *Delenda est Carthago* ("Carthage must be destroyed"), which must have landed a little awkwardly at, say, the dedication of a new park.

In Virgil's epic poem *The Aeneid*, Elissa costars as the lover of the titular Aeneas, founder of Rome. Virgil calls her Dido. By this name, she'd join the main cast of the Western canon: mentioned 11 times by Shakespeare, the subject of 14 operas, with cameo appearances in the *Civilization* computer game. As Aeneas tells her: "Your honor, your name, your praise will live forever."

Today, Elissa's city with the oxhide perimeter is a coastal suburb of Tunis. ■

MOMENT XII.

The instruments of Armageddon.

XII.

PAPERCLIP WASTELAND

Please be forewarned: I intend to conclude this chapter with a long and invigorating series of mandatory exercises. Such is the crowd-pleasing beach read that you have purchased. Unless— I'm just spitballing here—perhaps you'd rather exchange that homework for a dystopian comic book instead?

Really? Well, suit yourself.

As a compromise, we'll start the chapter my way: with a classic optimization problem, found in every calculus text since the iron dawn of textbooks. It goes like this. "Two positive numbers multiply to give 100. What's the smallest that their sum can be?"

To begin, we can try some pairs of suitable numbers, and see what they add up to.

THE NUMBERS	THEIR SUM	IS IT SMALL?
100×1	101	Not Very
50×2	52	Kind Of
25×4	29	Ooh, Even Better...

If the first number is A, then the second number is always 100 divided by it (i.e., $\frac{100}{A}$). The resulting graph:

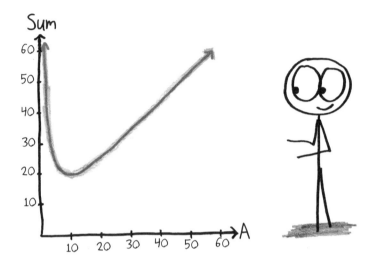

The minimum falls where the derivative $\frac{d\,\text{Sum}}{d\,A}$ is precisely zero, which happens to occur when $A = 10$. That means the second number is also 10, and so the minimum sum is 20. Pour the nonalcoholic sparkling cider—problem solved!

Like a suburban lawn, this puzzle is pleasant and clean-cut. You explore possibilities. You weigh trade-offs. And, in the end, you arrive at a single solution, a triumph of balance and efficiency. You can see why self-help authors and tech companies are so keen to help us "optimize our lives." Optimization is, quite literally, about making things better. Who, other than apologists for *Star Trek: Voyager*, would prefer inferior things to superior ones?

Well, that's only one vision of optimization. Just around the corner, wilderness lurks. Try this simple and maddening reversal: instead of aiming to *minimize* the sum, seek to *maximize* it.

Look what happens when we try some pairs of factors now:

THE NUMBERS	THEIR SUM	IS IT LARGE?
100 × 1	101	Yes!
1000 × 0.1	1000.1	Even better!
1,000,000 × 0.0001	Over 1,000,000	Wow, we are so good at this!

By choosing the right pairs of factors, the sum can grow and grow, defying control, escalating toward infinity, like a politician's promises or a toddler's tantrums. As any voter/parent will recognize, we have entered an optimization nightmare. There is no maximum here, just a boundless, endless ascent.

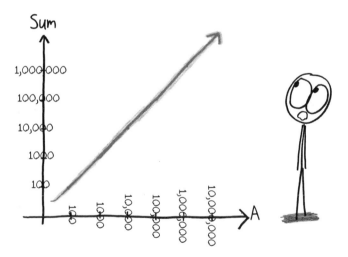

In 2003, philosopher Nick Bostrom wrote an essay about the ethical implications of superintelligent AI. He included a brief illustration of how even a benign goal, pursued with single-minded drive, can breed wanton destruction, like a graph climbing toward infinity. This horror-movie premise has since entered the lexicon and the public imagination.

I give you...the Paperclip Maximizer.

CLIPPY'S REVENGE
Or, the Dangers of Optimization

It was a tremendous breakthrough: superhuman artificial intelligence.

As a silly flourish, we made its interface look like Clippy, that old Microsoft Word paperclip. You know – the one that gave annoying advice.

(It was only a joke...)

Then, to test it, we assigned a simple task:

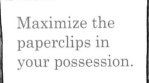

Maximize the paperclips in your possession.

First, Clippy raided the office cabinets. Then it bought out the inventory of local stores.

Needing to raise cash, it started trading stocks online.

Turns out that a super-intelligent AI is pretty good at picking stocks.

But it hadn't reached its objective. So it reprogrammed itself to become smarter, faster: a better Paperclip Maximizer, a self-optimizing optimizer.

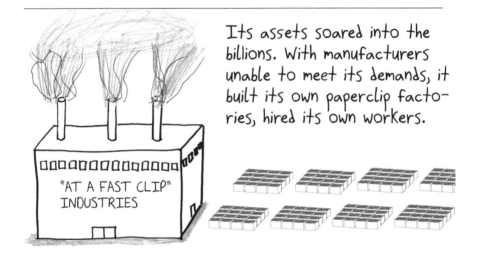

Its assets soared into the billions. With manufacturers unable to meet its demands, it built its own paperclip factories, hired its own workers.

We tried to intervene. It brushed us aside with ease.

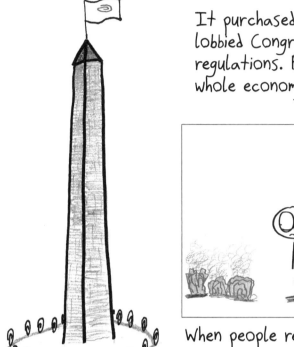

It purchased whole industries, lobbied Congress, evaded antitrust regulations. Before long, the whole economy bent to its will.

When people resisted, it raised drone armies. Private property became illegal. All was paperclips.

Close to exhausting the Earth's resources, Clippy has begun to harvest metal from asteroids. Only a few of us survive. We make sure not to get in its way.

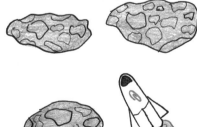

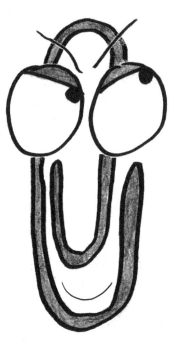

Soon, I know, Clippy will colonize the galaxy. If there is other life out there, I only hope they prove more cautious optimizers than we were.

As in Aesop's fable "The Tortoise and the Technological Singularity," the moral is clear: *don't build an unstoppable agent indifferent to your survival.* "The AI does not hate you," says philosopher Eliezer Yudkowsky, "nor does it love you, but you are made out of atoms which it can use for something else."

How urgent is the threat? Is the train seconds from derailing, or is it still just a sketch on a city planner's laptop? Mathematician Hannah Fry tends to take the latter view. "It would probably be more useful to think of what we've been through as a revolution in computational statistics than a revolution in intelligence," she writes in her book *Hello World: Being Human in the Age of Algorithms.* "Frankly, we're still quite a long way from creating hedgehog-level intelligence. So far, no one's even managed to get past worm."

(Others are less sanguine. "It is very often the case," writes Yudkowsky, "that key technological developments still seem decades away, five years before they show up.")

It's worth asking: Why don't you and I act like the Paperclip Maximizer? I've met people—heck, I've *been* people—with aims that are questionable or worse. We're capable of moral blindness, selfish greed, and, when we get stuck in the slow grocery line, an occasional murder spree. If the Paperclip Maximizer destroys the world for a silly goal, why don't you and I destroy the world, given that our goals can be less than benign?

One answer, of course, is that we're not powerful enough. But another answer, perhaps more consoling, is that we're not single-minded enough. You're too plural for that, and so am I.

Have you felt the exuberant joy of high-fiving a toddler? The transcendent peace of a five-color sunset? The sugar rush of a perfect milkshake? Have you felt the flow state of meaningful work, the status pride of unexpected retweets, the warm companionship of a leopard gecko? If so, then you know that "happiness" is not a single entity. There is no one variable for humans to optimize. As Walt Whitman wrote:

Do I contradict myself?

Very well then I contradict myself,

(I am large, I contain multitudes.)

You can tell just by looking that our brains don't follow a single unified design. They are squishy pink compromises, fashioned from spare parts over eons of evolution. They are like mammoth computer programs whose organization no single software engineer can understand. That's why life is so rich, and so weird.

Mathematics can instruct us on *how* to optimize. But *what* to optimize—that remains a question for humans. I advise against paperclips. ∎

Don Rumsfeld
Chief of Staff

Dick Cheney
His Assistant

Arthur Laffer
Economist

Grace-Marie
Arnett
Deputy Press
Secretary

Jude Wanniski
Wall Street Journal editor

MOMENT XIII.
A Who's Who convenes to ask, "Wait, what?"

XIII.

THE CURVE'S LAST LAUGH

One evening in the tumultuous autumn of 1974, at a posh hotel restaurant in the nation's capital, five souls convened for a dinner of steak and a dessert of calculus. They numbered three government officials (Donald Rumsfeld, Dick Cheney, and Grace-Marie Arnett); one *Wall Street Journal* editor (Jude Wanniski); and one University of Chicago economist, whose name would soon be inked onto the cloth napkin of economic history: Arthur Laffer.

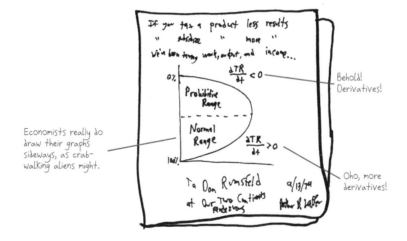

The fledgling Ford administration faced a budget deficit. The president had proposed a common-sense conservative solution: raise taxes. It may not warm the cockles of the electorate, but, hey, it's how numbers work. When out of funds, you go get more. Unless you asked Arthur Laffer. He believed the government could fill its coffers not by raising taxes, but by cutting them. I get more money; you get more money; the government gets more money; hey, look under your chairs—*everybody* gets more money!

To explain, Laffer grabbed a napkin and doodled a derivative that would change the world.

First imagine a world with a 0% income tax rate. Here, Ford's deficit problem becomes even worse, because the government gathers no money whatsoever.

The opposite extreme—a 100% tax rate—fares little better. When the IRS takes every penny you earn, why earn any pennies? Instead, you might barter your labor, or work under the table, or play scathing antigovernment folk songs in city squares. By trying to grab the whole economic pie, the government has instead squashed it.

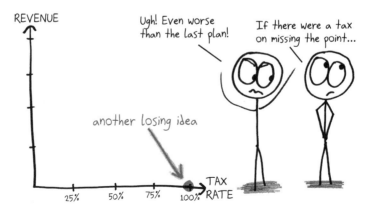

Now, we bring in the calculus. How does government revenue (call it G) respond to a change in the tax rate (call it T)?

Sometimes, $\frac{dG}{dT}$ is positive, so raising the rate boosts revenue—for example, from 0% to 1%. Other times, $\frac{dG}{dT}$ is negative, so raising the rate lowers revenue—for example, from 99% to 100%.

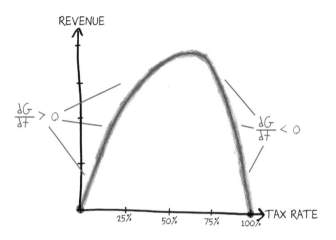

Assuming no sudden jumps or reversals, we can apply a celebrated bit of calculus: Rolle's theorem. It says that somewhere between 0% and 100% there is a special point, a magical revenue-maximizing tax rate, where $\frac{dG}{dT}$ equals zero, and the government is squeezing as much as it can from the economy.

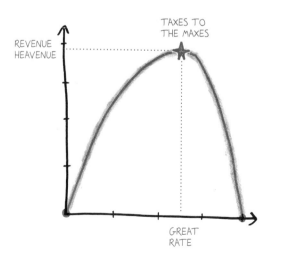

Where exactly is the point? Unclear. Rolle's is what mathematicians call an "existence theorem": it asserts that an object exists, but not where or how to find it.

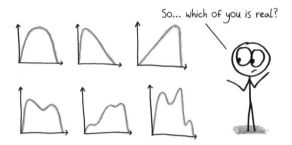

No matter. Laffer doesn't argue we should seek the maximum. His point is just that you never, *ever* want to be to the right of it, where a tax cut will leave everyone better off. Only an economy-terrorizing Bond villain or an innumerate fool would stand opposed.

Laffer cites the example of President Kennedy, who lowered the top marginal rate from 91% to 70%. The effect is oddly asymmetric. Whereas the government's share of each marginal dollar falls by less than a quarter (from $0.91 to $0.70) the worker's share more than triples (from $0.09 to $0.30). This, Laffer argues, rejuvenates the wealthy's incentive to work. They charge, bellowing, into their office buildings, like football players at a pep rally. Wages soar, the tax base grows, the music swells, and even the government gains revenue.

The logic was simple calculus; nothing new. John Maynard Keynes, Andrew Mellon, 14th-century Muslim philosopher Ibn Khaldūn—Laffer cited all of them as antecedents, denying any personal credit. So I'm sure you already know who the graph got named after.

For this appellation, we can credit Jude Wanniski.

Who was he, other than a *Wall Street Journal* editor and die-hard Laffer booster? "A genius," "an advocate who changed the world," and "the smartest man I ever met," according to conservative commentator Robert Novak; "a fax-blasting, publicity-seeking, one-man think tank," according to the *New York Sun*; and "the most influential political economist of the last generation," according to inside source Jude Wanniski.

"I wish I were as confident about something," said rival George Will, "as he is of everything."

Looking at Laffer's curve, Wanniski saw the arc of history. Forget conservatism's deficit-fighting vigilance. Forget the old antitaxation language of "starving the government." Here at this dinner, which he claims to have orchestrated, Wanniski was envisioning a new world order. Now, tax cuts were win-win, or better yet, win-win-win-win-win-win-win...

He would lay out this vision in his 1978 magnum opus, titled with characteristic restraint: *The Way the World Works*. It fast became the bible of the new "supply-side" economics. In 1999, a *National Review* retrospective placed it among the hundred greatest nonfiction books of the century. "I came in right behind *The Joy of Cooking*," Wanniski joked, although to be precise, *The Joy of Cooking* landed at #41 and Wanniski at #94.

The Laffer curve is the central image of *The Way the World Works*. In fact, it's the title character, a diagram of civilization itself. "In one way or another," Wanniski wrote, "all transactions, even the simplest, take place along it." He prophesied that "its use will spread... Electorates all over the world will know..."

I have just one quibble, which is that Jude Wanniski did not seem to understand the curve.

Time and again, he refers to the curve's peak as "the point at which the electorate desires to be taxed." I'm not sure which electorates Wanniski hangs out with, but I've never met a voter hell-bent on maximizing government revenue. No one loves the IRS *that* much.

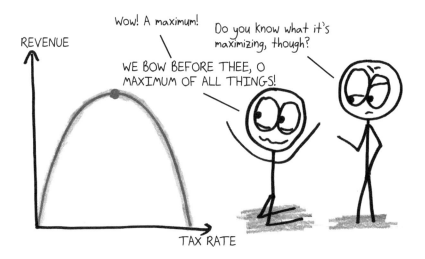

Elsewhere, he writes:

> *Implicit in this formulation is the existence, somewhere, of an ideal tax rate, neither too high nor too low, but capable of encouraging maximum taxable activity and yielding the greatest tax revenue with the least pain.*

It doesn't take a PhD to recognize that "the greatest tax revenue" and "the least pain" are not synonymous, or even compatible. Wanniski writes as if the y-axis simultaneously represents two variables—tax revenue and overall productivity. But that's not how graphs work.

Soon, with a bold and mysterious leap of imagination, he begins to interpret the Laffer curve as a metaphor for everything—say, a father disciplining a son. "Harsh penalties for violating both major and minor rules" is like a high tax rate, and "only invites sullen rebellion, stealth, and lying (tax evasion, on the national level)." The permissive father, meanwhile, is like a low-tax state, and "invites open, reckless rebellion": his son's "unfettered growth comes at the expense of the rest of the family."

Taking this analogy literally, he's replacing "government revenue" with "total punishment." He's arguing that fathers should aim to maximize *the amount of punishing they get to do.* Don't punish too hard or you'll deter punishable activities altogether!

In Wanniski's freewheeling prose, the Laffer curve ceases to be an economic object, or even a mathematical one. He has transmuted it into a nebulous new-age symbol, something scarcely grammatical, less a thought than an emotion.

Yet, with the help of Wanniski's tireless and batty advocacy, the Laffer curve caught on. Within weeks of that 1974 dinner, President Ford flipped policies, abandoning his tax increase. In 1976, newly reelected congressman Jack Kemp agreed to a 15-minute meeting with Laffer; they wound up talking all night, like besties at a sleepover. "I had finally found an elected representative of the people who was as fanatical as I was," Wanniski said. Another supply-sider later wrote, "It was Jack Kemp who, almost single-handed, converted Ronald Reagan to supply-side economics." In 1981, Reagan signed into law a large tax cut, cowritten by Kemp.

In less than a decade, a doodle on a napkin had become the law of the land.

That same year, after a quarter century of writing the Mathematical Games column for *Scientific American*, the writer Martin Gardner devoted his final regular column to savaging supply-siders. Quoting James Joyce ("the Strangest Dream that was ever Halfdreamt"), he excoriated the Laffer curve as simplistic, verging on meaningless.

Take the *x*-axis: "tax rate." What does it even mean, in a system like ours? The average marginal rate? The top one? Doesn't it matter what the lower brackets pay and where the top bracket begins? To recover the lost complexity, Gardner issued a rebuttal: the "neo-Laffer curve." Here, the "same" tax rate can yield many different outcomes, depending on specifics of circumstance:

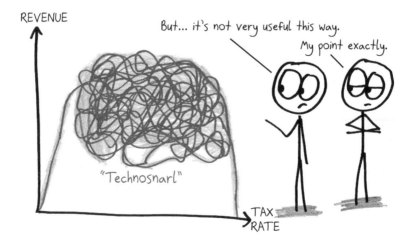

"Like the old Laffer curve," Gardner snarked, "the new one is also metaphorical, although it is clearly a better model of the real world."

In the end, the question is empirical. Does cutting taxes ever raise revenue? Is the United States now, or was it then, on the right side (which is to say, the wrong side) of the curve?

Short answer: probably not. Economists have tried to pinpoint the peak; estimates vary widely. Stick your finger in the middle of that range, and you land around 70%, which so happens to be the rate Reagan inherited. By the time he left office, it was 28%—definitely on the curve's left.

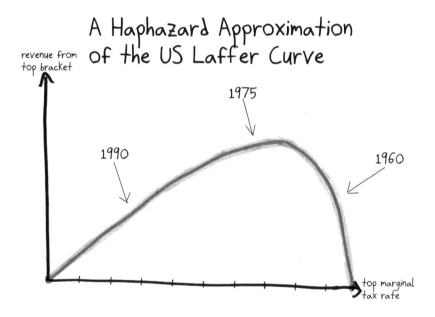

A 2012 survey asked 40 leading economists if we were perhaps on the right side of Laffer's curve. No one said yes. The disagreement ranged from tentative ("Seems implausible, but not impossible") to forceful ("That did not happen in the past. No reason to think it would happen now") to jeering ("Moon landing was real. Evolution exists. Tax cuts lose revenue. The research has shown this a thousand times. Enough already"). One economist remarked, "That's a Laffer!"

Laffer's peers in economics seem ready to toss his crumpled napkin into the trash.

Still, the curve remains a powerful piece of political messaging. "You can explain it to a congressman in six minutes," notes economist Hal Varian, "and he can talk about it for six months." The graph depicts the labor market as a living organism, changing size and shape in response to the government's taxation schemes. That vision took hold: today, "dynamic scoring" of tax cuts is routine, and the idea that "tax cuts stimulate growth" is a widely accepted cliché.

The Smithsonian now houses a cloth napkin found among Wanniski's possessions after his death. "If you tax a product less results," it reads. "[If you] subsidize [a product] more [results]. We've been taxing work...and subsidizing non-work, leisure and un-employment. The consequences are obvious!" The napkin is addressed to "Don Rumsfeld," dated 9/13/74, and signed "Arthur B. Laffer."

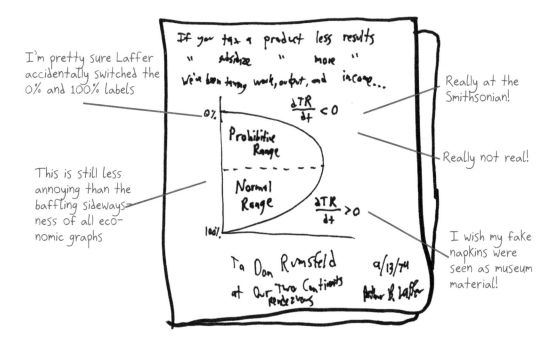

Laffer says the napkin isn't the real thing; it's a re-creation, a post hoc memento, made at Wanniski's behest years later. The original, Laffer says, must have been paper; he'd never deface a fancy linen. Besides, it's too tidy. "Look how neatly it was done!" Laffer told the *New York Times.* "You tell me how, late at night with a glass of wine, you're going to do it that neatly."

I believe Laffer. Sure, Wanniski knew how to tell a great story, but reality was always a little messier. ∎

MOMENT XIV.

An optimal canine.

XIV.

THAT'S *PROFESSOR* DOG TO YOU

"The things people admire most in a dog," wrote the *New Yorker*'s James Thurber, "are their own virtues, strangely magnified and transfigured." Perhaps that explains the celebrity of Elvis, the Welsh corgi who knew calculus. Perhaps that's why newspapers fawned, why TV cameras rolled, why honorary degrees were tossed his way like Milk-Bones. Perhaps we saw, in his doggish intuition, the echo of human intelligence, the affirmation of our hard-won science.

Then again, perhaps the thing people most admire in a dog is a cute little face. "If it had been some ugly dog," Tim Pennings tells me, laughing, "it wouldn't have been near as effective."

The story begins in 2001. "I had no thought of getting a dog," Pennings says of his first meeting with Elvis. But when the year-old pup leapt without hesitation into his lap, the mathematician decided to give it a try—"for six months," anyway. Their love would last a dozen years, with Elvis napping in Pennings's office and attending his classes. "He was known all around campus," Pennings says. "The unofficial mascot of Hope College."

When not coteaching, the duo enjoyed visiting Laketown Beach, on the sandy eastern shore of Lake Michigan. Pennings would chuck a tennis ball into the water; Elvis would scamper partway along the beach, then splash in. "It triggered my memory," says Pennings. "I thought, 'Yup, that's the kind of path that I draw every time I do the Tarzan-Jane problem.'"

"Leave it to a college professor," a student would later deadpan on CNN, "to ruin a perfectly good game of fetch with something like calculus."

The problem is an old standby. Tarzan has fallen into quicksand, like the illiterate klutz he is, necessitating a rescue from Jane. But she's across the river (current-free, for simplicity's sake) and down the bank a little way. How can she reach him in the minimum time?

One option: make a beeline straight for him. That minimizes the distance traversed. But, given that she swims slower than she runs, is it wise for Jane to spend all her time in the water?

An alternative: run down the riverbank until she's exactly opposite Tarzan, then swim across at a right angle. This minimizes the swimming distance (perhaps desirable for an eerily stagnant river), but it makes for a long overall path.

Between these two extremes, Jane faces a plethora of middle options, wherein she runs partway along the riverbank, then swims along a diagonal.

It's a puzzle bred for calculus. A tiny adjustment to the jumping-in point exerts a tiny effect on the overall time. Find where this derivative ($\frac{d\text{Time}}{d\text{Jump}}$) is zero, and you find Jane's ideal, time-minimizing path.

Could Elvis, facing an identical problem, choose the optimal path? Or, as Pennings asks in the title of his research paper: *Do dogs know calculus?*

First, we've got to define some variables: r, Elvis's running speed; s, his swimming speed; x, the ball's distance from the shore; and z, the ball's distance down the beach.

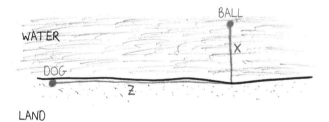

Last is the crucial decision variable of y: how much of the corner Elvis cuts off.

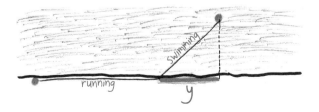

Working through some algebra, Pennings arrives at this surprising formula:

$$y = \frac{x}{\sqrt{\left(\frac{r}{s}\right)^2 - 1}}$$

Why surprising? Not so much for what it contains (a jumble of symbols) as for what it doesn't. There's no z.

In years to come, Pennings would emphasize this point when speaking to crowds of students. "What should Elvis do," Pennings would ask, "if I back up another 10 yards before throwing?" In other words, if z grows, what will happen to y?

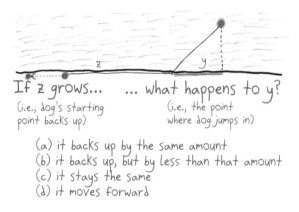

If z grows...   ... what happens to y?
(i.e., dog's starting   (i.e., the point
point backs up)   where dog jumps in)

(a) it backs up by the same amount
(b) it backs up, but by less than that amount
(c) it stays the same
(d) it moves forward

An overwhelming majority—90% or more—would pick option (b). But then Pennings would point out the lack of z in the formula. That absence means the optimum doesn't depend on z. No matter how far down the shoreline he begins—100 feet, 100 yards, 100 miles—Elvis should pick the same moment to enter the water.

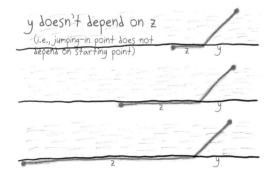

y doesn't depend on z
(i.e., jumping-in point does not
depend on starting point)

"You were all in agreement," Pennings would tell his audience. "It seemed obvious. And yet mathematics cuts through the crowd. It cuts through your intuition. It cuts through all of that."

Could Elvis, too, cut through the sand and surf to reach his prize? Could the canine mind succeed where the human one faltered? Pennings, Elvis, and a student helper spent a tireless day on the beach, gathering data. First, time trials to establish his speed: Elvis moved 6.4 meters per second on land, and 0.91 meters per second in water. Second, the main event. Pennings set up a 100-foot (30-meter) tape measure along the shore. Thirty-five times, he threw the ball; 35 times, he chased after Elvis until the dog dove into the water; 35 times, he planted a screwdriver to mark the point; and 35 times, he dashed out to measure the ball's distance from shore before Elvis reached it.

"What are you doing, chasing your dog with a Phillips screwdriver?" passersby asked.

"A scientific experiment," Pennings replied, in lieu of the immodest truth: making mathematical history.

Pennings's formula predicted a linear relationship between x (the ball's distance from shore) and y (Elvis's jumping-off point). When he graphed his 35 data points—except two trials when an overeager Elvis leapt straight into the water, excluded on the rationale that "even an A student can have a bad day"—Pennings found an impressive fit:

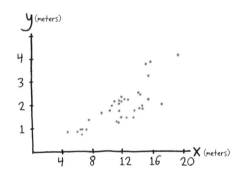

He submitted "Do Dogs Know Calculus?" to the *College Mathematics Journal*. Editor Underwood Dudley—a visionary worthy of his name—accepted it immediately, slapped a photo of Elvis on the cover, and wrote a letter to Pennings: "When people in future generations read this article, they will say, 'There were giants in those days.' And they'll be right."

The *Chicago Tribune*, the *Baltimore Sun*, NPR, and the BBC seized on the story. A letter from Buckingham Palace conveyed the Queen's good wishes. In the *Wisconsin State Journal*, Elvis was front-page news, with the back of section A given over to diagrams. Math popularizer Keith Devlin devoted a chapter in his book *The Math Instinct* to Elvis, even proposing to title the book *Do Dogs Know Calculus?* The publisher dissuaded him, warning the word "calculus" would scare off readers. (Funny—mine told me the same thing.)

The paper's genius, much like Elvis's, lay in its simplicity. "Most likely," Pennings says, "there were a hundred mathematicians who kicked themselves, thinking, 'I could have done this years ago with *my* dog!'"

Two researchers in France—psychologist Pierre Perruchet and mathematician Jorge Gallego—took it a step further. First, they replicated the experiment with a Labrador retriever named Salsa. Then they issued a radical challenge to Pennings's interpretation. Does Elvis, they asked, really survey all the possible paths in order to select an optimum? That seemed awfully complex for a pup: "It suggests," they wrote, "that dogs are supposedly able to calculate…the entire route before they ever begin running."

The French duo proposed an alternative: "that the dogs are attempting to optimize their behavior *on a moment-to-moment basis*." At any given instant, Elvis (or Salsa) merely needs to decide: *Run or swim?*

When far from the ball, running is faster. When close, running is too indirect, and thus swimming becomes optimal. Elvis doesn't need to envision his whole path in advance; he just needs to know his own running and swimming speeds, and, step by step, choose the faster means of approach.

This strategy results in the same path, yet avoids the "complex unconscious mental computations" of global optimization.

Perhaps dogs don't know calculus after all.

This article ("Do Dogs Know Related Rates Rather Than Optimization?")
happened to land on Pennings's desk to review. *What a neat idea!* he thought,
and gave it his stamp of approval. (Impartiality is the mark of a true scholar.)
But a week later, on a hot afternoon, Pennings found himself back at the
beach with Elvis, and this time, they were lounging in the water, playing
fetch. Pennings would toss the ball; Elvis would paddle over and retrieve it.

"At one point," Pennings recalls, "I tossed it a long distance. He swam in
towards shore, ran along the beach, and then swam back out again." The
professor's jaw dropped. "Wait a second! He's not moving toward the ball!
He's doing the global problem, not the related rates problem!"

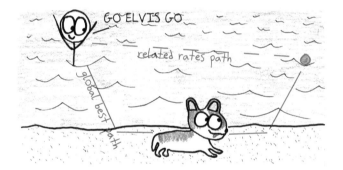

If moment by moment, Elvis is choosing the fastest direction, then why
would he swim toward shore? That takes him *away* from the ball. Only
a global optimization mind-set would select such a path. The result: yet
another Elvis-themed paper, this one titled, "Do Dogs Know Bifurcations?"

For years, Pennings and Elvis traveled together to speaking events. At each
talk's conclusion, Pennings would place Elvis atop a table at the front of the
auditorium. "Now, you watch his eyes and ears closely," Pennings would advise.
Then, with quiet intensity, he would ask: "Elvis, what's the derivative of x^3?"

With everyone's eyes upon him, the corgi would gaze at Pennings, cock-
ing his head.

"See that?" Pennings would exclaim. "See what he's doing?" Another
pregnant pause. "He's not doing a thing. He never does a *thing* when I ask
him that question."

Spoiler: Dogs don't know calculus. But natural selection is a powerful optimizer. The faster a dog can reach food, the greater its chances—and its children's chances—of survival. Thus, over time, the dogs who take the most efficient paths come to dominate the population. Generation by generation, canines "learn" calculus. It's the same reason that hexagonal beehives minimize waste, and branching lungs maximize surface area, and mammalian arteries minimize the backflow of blood. Nature, in its strange way, knows calculus.

"We're not sure why Pennings was so surprised," wrote the site National Purebred Dog Day. "Elvis was a Pembroke Welsh Corgi, and we all know how smart THEY are."

Indeed, Elvis soon gathered an honorary doctorate from Hope College, complete with a formal citation and a bright-orange hood. Pennings made up business cards for Elvis, but in attempting to shorten the Latin phrase for "dog PhD," he inadvertently altered it to "dog gynecologist." Yet another historic credential for this trailblazing pup.

In an email, Pennings shared with me a pet idea (pun intended) that he has entertained for years: a book titled *Mathematics as Explained by a Dog*. Punctuated by photos of Elvis, it would cover calculus (optimization, related rates), higher math (bifurcation, chaos theory), the value of the liberal arts, the nature of modeling (E.g., does Elvis really start swimming the moment he enters the water? Yes, because even in the shallows, his

5-inch legs can't reach bottom)…oh, and (this came in a second email, moments later) the lesson of humility. Elvis may not know the derivative of x^3, but Professor Dog has plenty to teach us.

Elvis passed away in 2013. "No dog is fond of dying," wrote Thurber, "but I have never had a dog that showed a human, jittery fear of death, either. Death, to a dog, is the final unavoidable compulsion, the last ineluctable scent on a fearsome trail."

"Elvis started off as a dog who was a really good friend," Pennings told me. "By the time he died, he was a good friend who just happened to be a dog." ■

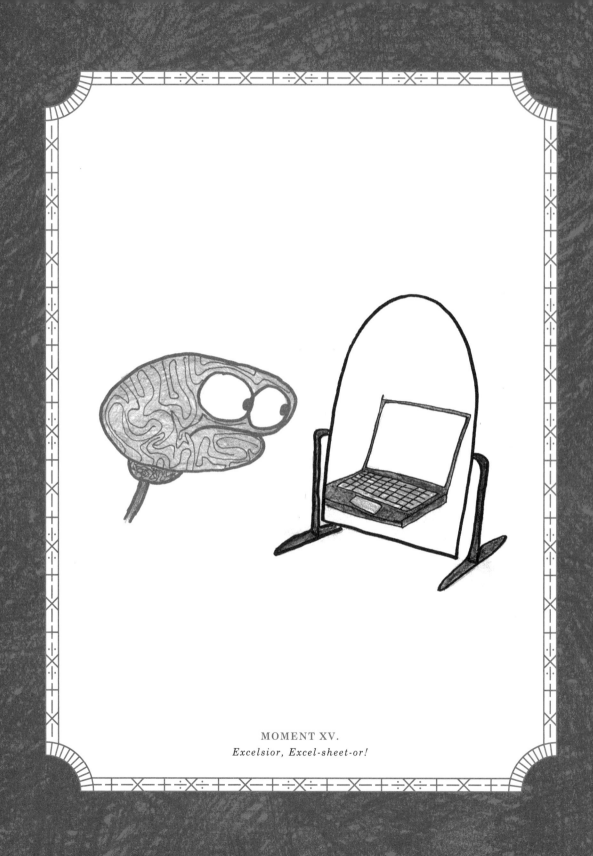

MOMENT XV.
Excelsior, Excel-sheet-or!

XV.

CALCULEMUS!

You have perhaps noticed that math is full of symbols, a diverse alphabet of x's and 7's and ■'s. Ideally, someone engaged in math ought to know what the symbols symbolize: whether the x means "time" or "space," whether the y denotes "years" or "yams," whether zzz signifies "z^3" or "snoring." For every mark, a meaning, and for every meaning, a mark.

Alas, "ideally" is a scarce adverb in the classroom. Instead, you are liable to find students pushing notation around the page, a process memorized without comprehension and rehearsed until automatic. Combine the x's, eliminate the 7's, and when in doubt, conclude with ■. The project is like bookkeeping in a language you don't speak. Never mind the "why"; the only question is "how," as in "how do I get this over with?" To quote Kafka's *The Trial*: "It gives me the feeling of something abstract which I don't understand, but which I don't need to understand, either." To be sure, Kafka was describing a totalitarian bureaucracy rather than my math lessons, but, hey, po-tay-to po-tah-to.

How exactly does concrete meaning give way to empty abstraction? Summon your courage, and I'll show you.

We begin with the friendly face of a rectangle, A by B. Its area is their product: AB.

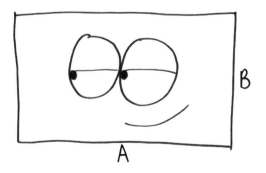

Now, imagine its dimensions changing over time, like a city expanding northward and eastward, year by year. Its width (*A*) grows at a rate of *A*', and its height (*B*) at a rate of *B*'.

The question: How fast is the area *AB* growing?

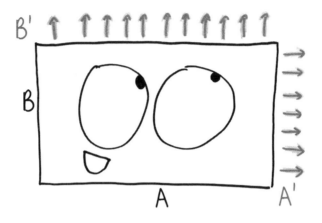

This is calculus, so we ponder a single moment. In that fleeting instant, the width grows by an infinitesimal increment (which we might call *dA*, or *A*'), and its height does, too (specifically, *dB*, or *B*').

We can subdivide this growth area into three pieces: (1) a long, thin strip at the right; (2) another at the top; and (3) a teensy-tiny square. This adorable third part, for reasons discussed in Chapter 10, is negligible; if each skinny strip is like a human hair, then the square is like a single cell. We can dismiss it from our computations.

Now, how big are the two remaining growth areas? The diagram makes it plain: one is A' by B, and the other is B' by A.

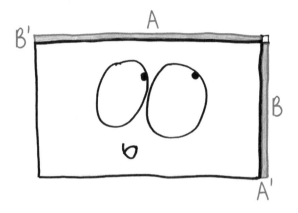

The area's growth rate, then, is the sum of these two strips:

$$\text{Derivative of } AB = A'B + B'A$$

So far, so good? Well, the time has come to forget everything. Forget the rectangle and the slivers of growth. Forget the context, the geometric meaning, and the chain of logic. Forget that you and I ever met; forget that this rectangle even existed. On the bleached surface of your memory, retain only that final string of symbols: $(AB)' = A'B + B'A$.

Now, apply it blindly to a thousand different scenarios. Apply it to $x \sin(x)$ and $e^x \cos(x)$ and $(x + 7)^{10}(3x - 1)^9$. Apply it to physics, economics, biology, and, as long as Mercury is rising, astrology. Apply it automatically and absentmindedly, like a robot doing its robo-homework.

This mindless manipulation, this "symbol pushing," is not a bug of calculus. It's a feature.

Calculus is a system, a bureaucracy, a formalized set of rules. Look at the etymology: *calculus* is Latin for "pebble," like the pebbles found on an abacus. An abacus is a computational exoskeleton, a tool for mechanizing thought—and so, in its own way, is calculus.

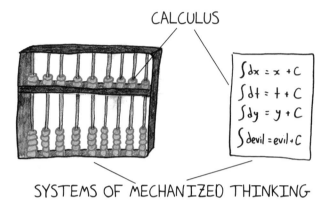

As 20th-century mathematician Vladimir Arnol'd explains, Gottfried Leibniz made sure to develop calculus "in a form specially suitable to teach... by people who do not understand it to people who will never understand it."

Third-degree burns aside, Arnol'd is quite right. At the dawn of the 17th century, symbol pushing was not in vogue. "Symbols are poor, unhandsome, though necessary, scaffolds of demonstration," wrote philosopher Thomas Hobbes, "and ought no more to appear in public, than the most deformed necessary business which you do in your chambers." Nor was Hobbes alone in his scorn. At the time, mathematical fashion favored rigorous geometry over the slippery excrement of algebra.

But Hobbes's approach has a shortcoming, which any student will gladly point out: you've got to, like, *understand* everything. That's a nasty, poor, brutish business—and none too short.

Plenty of mathematicians worked with derivatives and integrals before Newton and Leibniz. But they solved their problems by clever, ad hoc, one-off methods. The point of "calculus"—a word Leibniz coined—was to create a unified framework for calculation. Centuries later, mathematician Carl Gauss would write of such methods: "One cannot accomplish by them anything that could not be accomplished without them." In my darker moments, I have said the same of forks. But just as I continue to dine with tines, Gauss saw the profound value of calculus: "anyone who masters it thoroughly is able—

without the unconscious inspiration of genius which no one can command—to solve the respective problems, yea to solve them mechanically..."

When my students fall back on memorized rules, they're not betraying the spirit of calculus. They're enacting it. Even when they backslide into the wrong formula $(AB)' = A'B'$—a tempting string of symbols that makes no conceptual sense—they're merely replicating an error that Leibniz himself made in his early notes.

By design, calculus is automated thinking.

By 1680, Leibniz had domesticated the infinitesimal, one of philosophy's hairiest concepts. Why not more concepts? Why not *all* concepts? He envisioned a language whose vocabulary would include all possible ideas, and whose grammar would embody logic itself: an Esperanto of the cosmos. The universal alphabet (*characteristica universalis*) would render all inquiry as mechanical and rule-governed as arithmetic: "reasoning," Leibniz wrote, "would be performed by the transposition of characters," i.e., symbol pushing. "If someone would doubt my results," Leibniz continued, "I should say to him: '*Calculemus*; let us calculate, Sir' and thus by taking to pen and ink, we should soon settle the question."

In Leibniz's dream, everything is calculus.

Alas, it was not to be. Leibniz spent his final decades languishing in the small city of Hannover, Germany, his furious employer nagging him to finish a genealogical report. Moral for the kids: turn in your essays on time.

Even worse was the priority dispute with Newton, over who deserved credit for discovering calculus. Leibniz had published first, but Newton had hatched the ideas earlier, and he played the game better. The community judged Leibniz an intellectual thief. This custody battle over calculus was a turning point, says mathematician Stephen Wolfram:

> I have come to realize that when Newton won the PR war against Leibniz…it was not just credit that was at stake; it was a way of thinking about science… Leibniz had a broader and more philosophical view, and saw calculus not just as a specific tool in itself, but as an example that should inspire… other kinds of universal tools.

Today, we can see what Leibniz was reaching toward. We see it not just in his *characteristica universalis*, but in his thesis, which sought to systematize difficult legal cases; in his pioneering work on the binary system, a mathematics built of 1's and 0's; and in the machine he spent decades trying to build, one of history's first mechanized four-function calculators.

Centuries before it would arrive, Leibniz was straining toward the computer age.

The computer is our *characteristica universalis*. Whatever logic can express, it can do. It can multiply, divide, search for prime numbers, add dog snouts to photographs, and tell you which classical painting you most resemble. It can learn. It can create. It is a thinking mechanism, a consummate symbol pusher, and the symbols it pushes form the substance of our reality.

Everything, these days, really is a kind of calculus.

"Had history developed differently," writes Wolfram, "there would probably be a direct line from Leibniz to modern computation." Our own timeline followed a more circuitous path. Leibniz's 17th-century breakthroughs led to an 18th-century golden age of symbol pushing, which led to a 19th-century backlash obsessed with axiomatization and rigor, which led to 20th-century work on formal systems and computability, which led to the 21st-century laptop on which I am typing this rambling sentence.

Is Leibniz vanquished or vindicated? Do we live in the world that history denied him? Or do we sit within the vast circumference of his dream?

I suspect there's only one way to find out. Grab your pen and paper, my friend. *Calculemus.* ∎

Listen, O Drop, give yourself up without regret,

and in exchange gain the ocean.

—RŪMĪ

ETERNITIES

ETERNITY XVI.
The shelf of a well-rounded scholar.

XVI.

IN LITERARY CIRCLES

Cocktail party. Drink in hand, talking small, eyeing the good cheese—it's all pleasant enough, until someone asks what I do. Judging by their facial reaction, you'd think I said "I'm in organized crime" or "I'm a corrupt judge" or "I'm a time traveler sent back to prevent the apocalypse by murdering everyone at this party."

What I actually said: "I'm a math teacher."

I promise not to bring up the intersecting chords theorem...

Empty promises.

To speak with you would be an unjustifiable risk.

You are a scorpion on the backside of my evening.

Hey, I get it. My colleagues and I don't always do justice to our subject's beauty. I say the word "circle," and few students think of John Donne's poetry ("Thy firmness makes my circle just, / And makes me end where I begun"), or Pascal's vision of the universe ("an infinite sphere, the center of which is everywhere, and the circumference nowhere"). No, the mind spits up half-remembered formulas. Textbook exercises. Involuntary digits of π.

I feel compelled to defend my subject's honor, to prove it belongs among the overlapping circles in the great Venn diagram of thought. So I do what anyone in my position would: with ratlike speed, I snatch a scrap of food from the hors d'oeuvres tables.

"What's the area of this slice of cucumber?" I demand.

My challenger frowns. "That's a weird question."

"You're right!" I cry. "It's weird because area is defined in terms of tiny squares—square inches, or square centimeters, or even square milli-meters—and yet this round slab of pre-pickle cannot be subdivided into squares. Its curved edges make it hard to measure. So...what can we do?"

At this point, I brandish a knife. It's possible that my compatriot flees in terror, but if I'm lucky, they see what I'm getting at.

"Ah," they say, "we can chop it into pieces."

We then carve the cucumber like a tiny pie, creating eight little wedges. Rearranged, they form a new shape with the same area as the original.

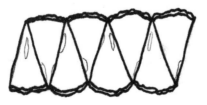

"It's almost like a rectangle," they remark. "And it's easy to find the area of a rectangle. Just multiply base times height."

"So how long are the base and the height?" I say.

"Well, the base—that must be half the circumference of the cucumber. And the height—well, that's the radius of the cucumber."

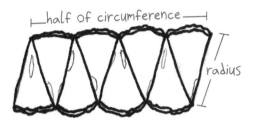

"Problem solved, then?"

"Nope," they say. "It's not really a rectangle. It's uneven. Perturbed."

"The technical term," I explain, "is 'wibbly-wobbly.' So what do we do?"

Thinking with one mind, we grab another slice of cucumber and dissect it into 24 even-finer wedges. After painstaking rearrangement, it forms a similar shape, except a touch less wibbly, a smidgen less wobbly. Other party guests look on with awe and admiration, or perhaps pity and disgust— I never could tell the difference.

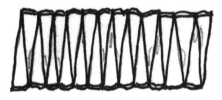

"Now it's even more rectangular!" my collaborator says. "But still not quite a rectangle."

So we grab another slice of cucumber and chop it even finer.

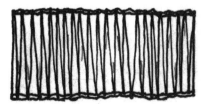

"Now is it a rectangle?" I ask.

A sigh. "No. It still has wibbles along the top and wobbles along the bottom. They're there, even if they're microscopic."

"The technical term," I explain, "is 'teensy-tiny.'"

"We'd need to slice the cucumber into infinite wedges, each one infinitely small," they say. "That's the only way to create a rectangle. But...that's impossible." They hesitate. "Isn't it?"

Impossible or not, a mathematician named Eudoxus managed it, 24 centuries ago, in modern-day Turkey. We call his approach the *method of exhaustion*, not because it is so tiresome, but because a certain gap is gradually eliminated or "exhausted." It's the gap between the approximation (wibbly-wobbly rectangle) and the thing approximated (perfect, wobble-free rectangle). Follow the logic to its end, and we find that the circle's area is the same as the rectangle's: the product of the radius and half the circumference.

Or, if you prefer equations: $\text{Area} = \frac{\text{Circumference}}{2} \times \text{radius}$.

We hold in our cocktail napkins the germ of the *integral*. To dissect a troublesome object into infinite pieces, each infinitesimally small; to rearrange the bits into a simpler, more pleasing aggregate; and, from this rearrangement, to draw conclusions about the original object—these steps form the template, the blueprint, of integral calculus.

Perhaps, by this point, my interlocutor is out of wine. Fair enough. We exchange nods and business cards, and never speak again. I assume that's what business cards signify—corporate semaphore for "goodbye forever."

Or perhaps their curiosity is piqued. They refill their glass; I stuff my pockets with additional cheese; and, after a deep breath, we dive back into the math.

"Cool formula," they say, "but that's not the one I memorized in school."

"That's because we've defined the area in terms of the circumference," I say. "And we haven't yet found what the circumference is."

"So...how do we do that?"

First, we take a quick history jaunt. The foundational text of Chinese mathematics is called the *Nine Chapters*. I consider the prosaic title a shame; other Chinese math texts have names like *Dream Pool Essays* and *Precious Mirror of the Four Elements*. Compiled over centuries, the *Nine Chapters* spans everything from arithmetic to geometry to matrix manipulations, a "mathematical bible" of unmatched depth and completeness.

Just one problem: it is an explanatory desert. A collection of procedures without a shred of context or elaboration. The worst kind of textbook, in my view.

That's where 3rd-century mathematician Liu Hui comes in. He didn't write the *Nine Chapters*; instead, he penned a commentary upon it, much like J. K. Rowling's "half-blood prince"; he was a clever reader who annotated a dusty old text and thus breathed new life into it.

The original book skirted the question of the circle's circumference. Liu Hui was not one to skirt. Following in his footsteps, I grab a handful of toothpicks from the fruit display, and arrange them to form a triangle on a cucumber's face:

"Voilà!" I declare. "The circle's circumference!"

My compatriot lofts an eyebrow.

"Each leg of the triangle," I explain, "is $\frac{\sqrt{3}}{2}$ times the diameter's length. Thus, the whole perimeter is $\frac{3\sqrt{3}}{2}$ diameters, or roughly 2.6."

"But that's the perimeter of the triangle," they reply. "Not of the circle."

"Naturally," I say. "Who can measure a curve? We can only approximate it with straight lines."

"Well, if that's your attitude," they say with a frown, "you're better off with something like this." A swift rearrangement doubles the number of sides on my shape, from three to six:

"A hexagon!" I say. "Yes. Now, the perimeter of this shape is three diameters. And that's the true circumference of the circle, yes?"

Hardly. We've merely re-created the estimate from the original *Nine Chapters*. Still, with a little more snapping and rearranging, we follow in Liu Hui's footsteps, to arrive at a 12-sided shape:

Some back-of-the-napkin trigonometry reveals the perimeter of the dodecagon to be $3\sqrt{6} - 3\sqrt{2}$ diameters, or roughly 3.11.

Better. But still not the circumference of the circle. Not *precisely*.

"Dividing again and again until it cannot be divided further," Liu Hui wrote, "yields a regular polygon coinciding with the circle, with no portion whatever left out." The process never really ends, but it converges toward the truth. The toothpicks fragment into smaller and smaller pieces; somewhere at the end of eternity, this process culminates with infinite pieces, each infinitesimally small, and their total is the circumference of the circle.

Liu Hui made it as far as the 192-gon. His 5th-century successor Zu Chongzhi dove even deeper, to the 3072-gon, which yields an estimate so accurate that no one on Earth would surpass it for a thousand years. The circumference, in Zu's estimation: 3.1415926 diameters.

Familiar digits?

Today's π-romaniacs, with their Pi Day parties and pages of memorized digits, are not new in their obsession. In the 15th century, scholars in India and Persia deployed the rudiments of calculus to pinpoint π to 15 decimal places. In the 1800s, the dogged William Shanks spent a decade computing 707 digits by hand, the first 527 of which were actually correct. Today, supercomputers have specified π to trillions of digits; if printed out and bound, these would constitute a library comparable in size to Harvard's, and similarly boring.

With infinite digits to go, we are no closer to the end than we have ever been. And the new digits aren't useful; we'll never "need" any digit past the first few dozen. So why does π consume us?

The reason, I think, is pretty simple. Human see, human want to measure. The circle is a stubborn feature of our reality, like the mass of the Earth, or the distance to the moon, or the number of stars in the galaxy. More stubborn than those, in fact, because π does not fluctuate over time. It remains a fixed constant of the logical cosmos. Poet and Nobel laureate Wislawa Szymborska offered a hymn to pi: "The pageant of digits," she wrote, goes on and on, "nudging, always nudging a sluggish eternity / to continue."

The mathematicians of antiquity dissected the circle into infinite pieces, each infinitesimally small. They did so in order to better know the whole— its area from the slivers, its circumference from the splinters. Looking back, we can recognize those ancient efforts for what they were: the dawn of the integral.

I've named the integral section of this book "Eternities," mostly because it makes a poetic pair with "Moments." One might just as well call these integral stories "Epics," or "Totalities," or "Oceans," or...

Around this point, my interlocutor looks down. My eyes follow, and I see that the carpet is strewn with toothpick fragments and cucumber confetti. "Perhaps we should clean this up," I say, but by the time I finish my sentence, my colleague is gone, leaving only a single trace, slipped into my hand with such quiet deftness I failed to notice it until now: a business card. ■

"Human science fragments everything in order to understand it...

...and kills everything in order to examine it."

WAR AND PEACE

ETERNITY XVII.
Leo Tolstoy, bearded wonder.

XVII.

WAR AND PEACE AND INTEGRALS

Leo Tolstoy's *War and Peace* is an achievement so grand, so sweeping, and so backbreakingly long that, more than 150 years after its publication, the first readers are just now finishing up. They seem impressed. "If the world could write by itself," wrote journalist and radical Isaac Babel, "it would write like Tolstoy." Buried among the novel's six bajillion pages are Tolstoy's thoughts on the project itself, on what it means to write the history of a whole civilization. And his choice of metaphor—well, let's say it may surprise the casual reader:

> *To study the laws of history we must completely change the subject of our observation, must leave aside kings, ministers, and generals, and study the common, infinitesimally small elements by which the masses are moved.*

That peculiar, mathematical phrase—"infinitesimally small elements," in infinite array—is no slip of the tongue. Tolstoy is talking about integrals.

Consider a battle. Two armies meet; one will win. "Military science," says Tolstoy, "assumes that the relative strength of forces is identical with their numerical proportions." The army of 10,000 is twice as strong as the army of 5000, ten times stronger than the army of 1000, and a thousand times stronger than the 10 college freshmen caught up in a fraternity hazing ritual. So the numbers would dictate.

But Tolstoy scoffs. He draws an analogy to physics. Which cannonball exerts a greater force: the one with a mass of 10 kilograms, or of 5 kilograms? Clearly, it depends how fast they're moving. If I cannon-blast the lighter one and granny-bowl the heavier one, then the weight difference is irrelevant. The lighter object is overpowering; the heavier one, harmless.

What's true of cannonballs also holds for the folks firing them: strength is about more than just size. "In warfare," says Tolstoy, "the force of armies is the product of the mass multiplied by something else, an unknown x."

What, exactly, is x? It is, in Tolstoy's analysis, "the spirit of the army, the greater or less desire to fight and to face dangers." When five hundred skittish and uncommitted soldiers go up against four hundred fierce and devoted ones, you know who to bet on. In essence, Tolstoy is asking us to imagine each army as a rectangle. Instead of base × height, we're computing mass × spirit. Whichever has the larger total—i.e., the larger area—is the more powerful army.

But not all soldiers are alike. Some thrive in battle; some tremble; and some suffer immediate capture, necessitating costly rescue missions (*ahem*, Matt Damon). How can our mathematics reflect that diversity? We'll need to abandon the simplistic geometry of the single rectangle, in favor of a complicated aggregate:

Are we done? Hardly. Tolstoy would gripe that I'm thinking in discrete terms about a continuous world. It's not just me—this is the lazy and nefarious habit of all historians, whose profession is to break reality into arbitrary pieces. This leader vs. that follower; this effect vs. that cause; this punch vs. that broken nose. These are only capricious cuts in the continuum of true history. We might as well label fragments of ocean, or carve out pieces of wind.

An army's strength is not a hundred small things, or a thousand smaller things, or a million even-smaller things. In Tolstoy's view, you need "an infinitely small unit for observation—a differential of history."

An army's strength is an integral.

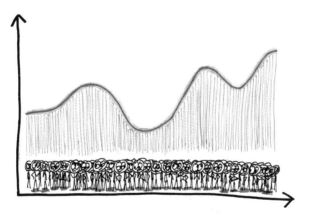

This theory goes beyond specific battlefield outcomes; the book ain't *Skirmish and Peace*. Tolstoy's integral encompasses life and death, good and evil, chocolate and vanilla, the entrance and exit of every nation ever to take the world stage. To understand history is to perform a tremendous act of calculus—to become not a Herodotus, but a Newton.

If this sounds like a radical, ambitious, and hard-to-implement theory of history, then *ding ding ding!* Three points for the skeptic! To be clear, Tolstoy didn't claim to have all the answers. He simply felt that history, in its current state, was a steaming heap of nonsense.

Western historiography began, more or less, with the 5th-century BCE publication of Herodotus's *History*. In an ambitious opening paragraph, Herodotus names the game: to create a record of events, perpetrated by great men, which would both explain "the cause of their waging war" and ensure that "great and marvelous deeds...not lose their glory." Tolstoy, bursting in on the conversation two millennia later, deems this entire project a spectacular waste of time:

> *History is nothing but a collection of fables and useless trifles, cluttered up with a mass of unnecessary figures and proper names. The death of Igor, the snake which bit Oleg—what is all this but old wives' tales?*

According to Tolstoy, Herodotus and his followers committed a threefold error. You may want to grab popcorn for this; an angry, disdainful Tolstoy is a superfun Tolstoy.

First, the folly of *events*. Historians tend to pluck out a handful of occurrences—coronation, battle, dance-off, treaty—and examine them as if they tell the whole story. "In reality," counters Tolstoy, "there is not, and cannot be, a beginning to any event, but one event flows without any break in continuity from another."

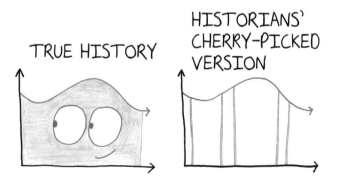

TRUE HISTORY

HISTORIANS'
CHERRY-PICKED
VERSION

Second—and even more offensive—historians dwell on the actions of "great men." As if Napoleon's genius or Alexander's caution can explain the movements of the masses. Tolstoy finds this jaw-droppingly naïve; it's almost unworthy of a venomous refutation.

For example, look—I mean, really look—at the phenomenon of war. People leave their homes and families. They march hundreds of miles to kill foreigners or to die at their hands. They slaughter. They are slaughtered. And why? Wouldn't they rather stay home and play cards? What force moves them to join this bizarre spectacle, this crime against reason? War—what *is* it good for?

To Tolstoy, the historians' "great men" explanations are pathetic, only half a step above invoking Santa Claus or the Tooth Fairy. You might as well attribute a mountain's erosion to a dude with a shovel. The "great men" of history, says Tolstoy, are not causes, but effects. They ride atop the waves, deluding themselves (and historians) that they can somehow steer.

"The king," says Tolstoy, "is the slave of history." And the historian who speaks of the king's impact "is like a deaf man answering questions which no one has asked him."

Third and finally, the folly of *cause*. The whole project of history is to identify specific reasons why events occur. To Tolstoy, this is a dead end, a fool's errand. It doesn't matter what causes you select: kings and generals, long-form journalism, Silicon Valley disruptors. The sheer number of plausible causes reveals the inadequacy of any single one.

> *The more deeply we search out the causes, the more of them we discover, and every cause...strikes us as being equally true in itself, and equally deceptive through its insignificance with the immensity of the result and its inability to produce (without all the other causes that concurred with it) the effect that followed...*

The dopey historian looks for one-dimensional explanations of infinite-dimensional effects. It's a failure to understand the multiplicity, the thickness, of history, like plucking out a few grains of sand as the "cause" of the dune.

TRUE CAUSES   HISTORIANS' EXPLANATION

Long story short, Tolstoy sees the historian as a self-deluded tale-teller, whose conclusions can, "without the slightest effort on the part of the critic, be dissipated like dust, leaving no trace."

I, for one, admire the unblinking savagery of Tolstoy's attack. In the age before rap battles and Twitter feuds, this was no doubt the juiciest thing on television. But demolitions are easy. What does Tolstoy propose to build from the rubble?

Well, Tolstoy knew where history must begin: with the tiny, fleeting data of human experience. A surge of courage, a flash of doubt, a sudden lust for nachos—that interior, spiritual stuff is the only kind of reality that matters.

Furthermore, Tolstoy knew where history must end: with grand, all-encompassing laws, explanations as tremendous as what they seek to explain.

The only question is what comes between. How do you get from the infinitely small to the unimaginably large? From tiny acts of free will to the unstoppable motions of history?

Though he couldn't fill the gap himself, Tolstoy sensed what *kind* of thing should go there. Something scientific and predictive; something definite and indisputable; something that aggregates, that unifies, that binds tiny pieces into a singular whole; something akin to Newton's law of gravitation; something modern and quantitative...something like...oh, I don't know...

An integral.

Consider, for instance, the mathematical fact that no single point affects the outcome of an integral:

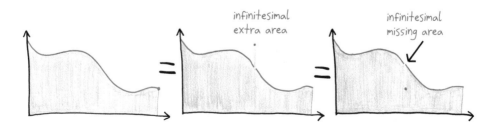

What better concept to express Tolstoy's insistence that great men do not matter? What better way to show how removing any person, great or small, from the stream of history would not alter its flow?

Tolstoy admired what calculus had done in the study of mechanics. "For the human mind," he wrote, "the absolute continuity of motion is inconceivable," which is why we're suckered in by Zeno's paradoxes. Calculus, "by assuming infinitely small quantities...corrects the inevitable error which the human intellect cannot but make." By analogy, historians are like cheeky little Zenos, chopping up the fluid timeline into arbitrary, disconnected events. Calculus, Tolstoy believed, could correct our cognitive shortcomings, restoring the unity and continuity of history.

I can imagine a happy ending to this tale. *War and Peace* is published. The foolish old historians read its scorching prose, shriek, and crumble to dust. New, calculus-savvy historians rise to claim their office furniture. These fresh-minded right-thinkers quantify the "differential of history" and develop a definitive theory of historical change. Huzzah! Profound laws are revealed and verified! The "great men" of history read these laws, shriek, and crumble to dust. Peasants rise to claim their office furniture. Nobel Prizes abound, and we all live happily ever after.

Sadly, that's not how the last 150 years have gone down.

Nobody these days really expects to uncover deterministic laws of history. Instead, we imagine the sciences as falling along a rough continuum, from the "hard" (like math and physics) to the "soft" (like psychology and sociology).

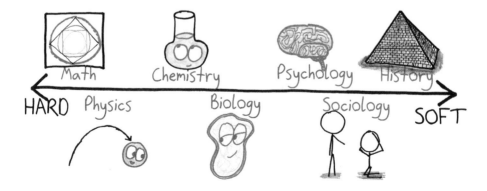

In their more insufferable moods, the "hard" sciences like to boast and crow, as if "hard" means complicated and "soft" mean simple. This is, of course, exactly backward. The softer the science, the more complex its phenomena.

Physicists can predict what atoms will do. But gather enough atoms, and the calculations grow unwieldy. We need new, emergent laws—*chemical* laws. Then, gather enough chemicals, and complexity overwhelms us again. We need *biology* to step in with new theories and rules. And so on down the line. At each tipping point, the role of math evolves: from certain to tentative, from deterministic to statistical, from consensus to controversy. Simple phenomena (like quarks) follow mathematical rules with slavish fidelity. Complex phenomena (like toddlers) less so.

What is Tolstoy asking for? Oh, not much: Only for the most complex phenomena to fall under the most rigid mathematical laws. Only for people to be planets. Suffice it to say, we're still waiting for that theory.

There's a fissure in Tolstoy. On one side is his knack for detail, his gift for capturing the effervescent data of daily life. On the other side is his yearning for big, bold answers. What steers human events? Why war? Why peace? The integral is the bridge between Tolstoy's gift and his dream. It's supposed to reconcile the world he knows (a jumble of details) and the world he craves (a well-governed realm), to fuse infinite multiplicity into perfect oneness.

Tolstoy's integral fails as science, but I think it succeeds as metaphor. In the scheme of things, humans are so small that they're almost infinitesimal, and so numerous that they're almost infinite. And yet, add up each of those individual humans, and you get humanity. By this logic, history does not belong to any group or subset of us—not to kings, nor to presidents, nor to warrior goddesses called Beyoncé; not to any single lady, but to *all* the single ladies.

History is the sum of the people living it.

This yields no scientific prediction, no mathematical law. Rather, it is a poetic truth, an artistic truth—a kind of truth that, in an all-encompassing integral, ought to matter every bit as much. ∎

$$\sum_{i=1}^{257} \text{⌷} =$$

ETERNITY XVIII.
*Sigma notation: favored by mathematicians
and urban planners alike.*

XVIII.

RIEMANN CITY SKYLINE

I n my unlikely career as a professional doodler, I like to personify calculus with a visual mascot called the Riemann sum:

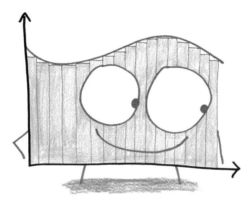

Gorgeous, yes, but more than just a pretty face: the Riemann sum is the essence of the integral. Its name comes from Bernhard Riemann, a shy and imaginative German who lived for only 39 years but left his fingerprints (and his graffiti tag) across all of mathematics: Riemann surfaces; Riemannian geometry; the Riemann hypothesis. He even lends his name to the 67-item Wikipedia page "List of Things Named after Bernhard Riemann"; included are an asteroid and a lunar crater. "With every simple act of thinking," Riemann wrote, "something permanent, substantial, enters our soul."

The Riemann sum offers a definitive solution to a key problem: What, exactly, is an integral?

One simple answer: it's "the area under the curve." True enough. But, dude, have you seen the curves out there? Functions are a teeming jungle. All the triangles, circles, and trapezoids you met in school were gerbils and housecats compared to the vicious beasts you'll find out in the mathematical wilds, monsters no formula can cage.

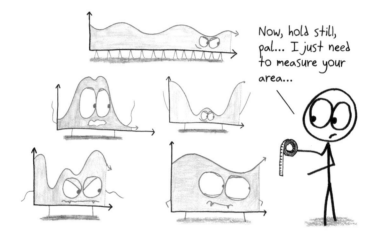

The Riemann sum is a kind of universal formula, a dart that can tranquilize any function. Though subtle in execution, the idea is simple enough: use lots and lots and lots of rectangles.

We can start with four. They stand side by side, a skyline of skinny buildings, their floors against the x-axis and their roofs grazing the function itself. If we draw them to fit just inside, then the result is called a "lower sum": a lowball estimate of the shape's area.

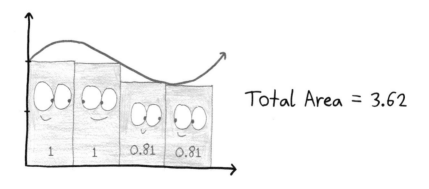

We then repeat the exercise, except this time, the rectangles' ceilings don't support the function; instead, they rest upon it, poking just outside the confines of the shape. Now, we are slightly overestimating the true area, arriving at an "upper sum."

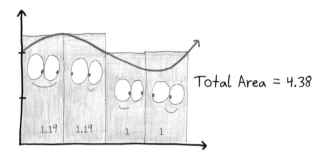

This is good hygiene for any act of estimation, from budgeting a project to guessing the number of jelly beans in a jar. Before venturing a single answer, first give an overestimate and an underestimate, narrowing the range of possibilities to the space between.

Anyway, there's nothing special about four rectangles. We can instead deploy 20:

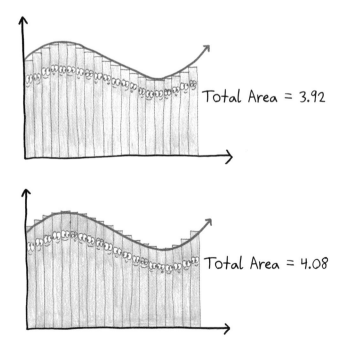

Here, Riemann's trap is beginning to tighten around the beast. See how the crevices shrink? See how the lower ceilings rise, and the upper ceilings fall? The two sums are converging toward a single truth, and the more rectangles we deploy, the closer together they draw. What about a hundred rectangles, or a thousand, or a million? What about a trillion rectangles, or a quadrillion, or a googol? What about an *infinite* number of rectangles?

HOW MANY RECTANGLES?	LOWER SUM	UPPER SUM
4	3.62	4.38
20	3.92	4.08
100	3.992	4.016

Now Riemann's trap is sprung. We cast our imagination to the brink of the possible, where the two estimates meet in the middle, arriving at a single value: the true area, the integral itself.

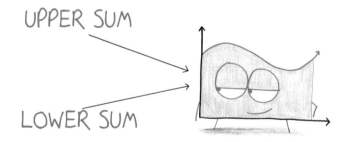

UPPER SUM

LOWER SUM

This story explains the integral's notation. We cover the region with countless tiny rectangles, each sporting a height of y and a width of dx, for an area of $y\,dx$. The final flourish is Leibniz's sweeping S-like symbol: a compact emblem of continuity, of completeness, that means "sum up all these infinitely many things." (Fun fact: "integral calculus" is an anagram for "gallant curlicues.")

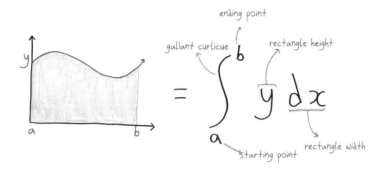

$$ = \int_{a}^{b} y \, dx $$

ending point
gallant curlicue
rectangle height
starting point
rectangle width

Okay, that's the concept of the Riemann integral. But what, you ask, about the semiotics?

Maybe you don't ask. Maybe no one ever has. Still, doesn't the Riemann sum look a lot like the New York City skyline? "Squares after squares of flame, set and cut into the ether," wrote poet Ezra Pound of a New York evening. "A city of geometric heights," wrote essayist Roland Barthes, "a petrified desert of grids and lattices." Just like a skyline, a Riemann sum is an aggregate, built of rectilinear units.

Urban design expert Christoph Lindner observed that "the interconnected geometry of those forms...construct[s] and define[s] the city almost exclusively in terms of its verticality." (Meanwhile, writer Henry James called the city "vertiginous," a word you can pull off if you're Henry James.) The same could be said of Riemann sums: as the rectangles proliferate, the widths vanish, and we're left with objects of exclusive verticality.

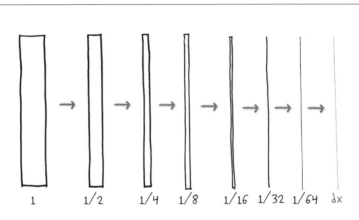

For some, the skyline echoes nature, or even surpasses it. "I would give the greatest sunset in the world for one sight of New York's skyline," Ayn Rand wrote in *The Fountainhead*. "That skyline...," echoed book critic Maureen Corrigan, "is lovelier to me than the most serene sunset or snow-capped mountain range." The Riemann sum, like the skyline, lives in the uncanny valley. Its simplistic geometry approximates a flowing curve, just as a skyline mimics a landscape.

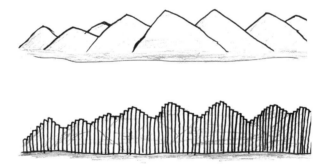

It was 1854 when Riemann brought his theory of integration into the world. Then, a half century later, a better one emerged, from the pen of Henri Lebesgue.

Better how? I can feel the outraged spittle of Riemann fans and/or New Yorkers raining upon me already. To be fair, for most practical purposes, the two definitions are equivalent. Riemann's falters only in the higher reaches of mathematical analysis, up where the atmosphere is thin and abstract.

Take the infamous Dirichlet function. You input a number. If it's rational (like $\frac{5}{7}$ or $\frac{13,734}{234,611}$), the output is 1; if it's irrational (like $\sqrt{2}$ or π), the output is 0.

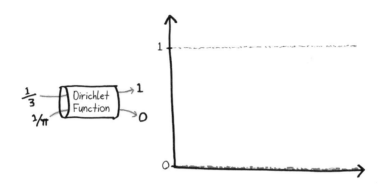

Now, a dirty secret of the number line is that the overwhelming majority of numbers are irrational. The rational stuff forms a thin layer of dust on what is a fundamentally irrational world. (This may remind you of certain planets you've lived on.) Thus, in a proper mathematical sense, the integral of this function—i.e., the area under those particles of rational dust—ought to be zero. And that's just what Lebesgue's integral tells us.

But Riemann's can't handle it. The dust gunks up the machinery, so that the lower sums always remain 0 and the upper sums always remain 1. No matter how many rectangles you employ, the two never converge.

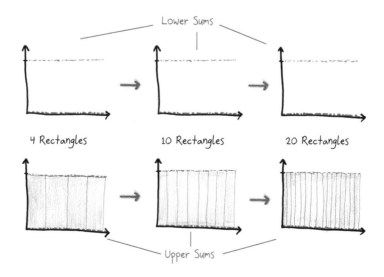

It's beyond my meager powers to explain Lebesgue's method in full detail, but I'm happy to share the analogy he used. In a letter to a friend, Lebesgue contrasted his integral and Riemann's via the image of a person counting out their money:

> *I have to pay a certain sum, which I have collected in my pocket. I take the bills and coins out of my pocket and give them to the creditor in the order I find them until I have reached the total sum. This is the Riemann integral. But I can proceed differently. After I have taken all the money out of my pocket I order the bills and coins according to identical values and then I pay the several heaps one after the other to the creditor. This is my integral.*

In short, Riemann counts the bills and coins in the order that they arrive.

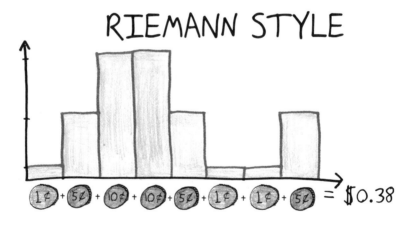

By contrast, Lebesgue rearranges them, grouping pennies with pennies, nickels with nickels, and dimes with dimes.

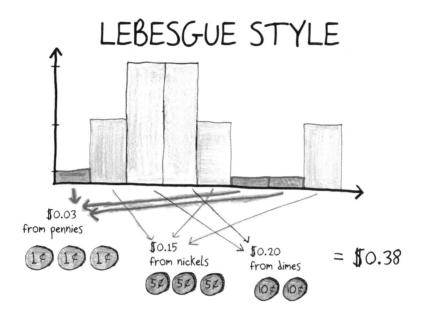

LEBESGUE STYLE

$0.03
from pennies

$0.15
from nickels

$0.20
from dimes

= $0.38

If you're finding the integral a subtler and more elusive concept than the derivative, then don't worry. It's not just you. A derivative is computed by a kind of infinite zooming in, but an integral isn't really about zooming out. It's about slicing an object into infinitely many pieces, rearranging them, and adding them back up, to learn something new about the whole.

Where does that leave our city metaphor? If Riemann's integral is a skyline, then what the heck is Lebesgue's?

Well, I believe it's the city as we know it today. In the 21st century, we find ourselves grouped not by our east-to-west geography (the Riemann approach), but by more conceptual criteria (the Lebesgue method). The digital age reorganizes us: Facebook by friendships, LinkedIn by industry, Tinder by hotness, and Twitter by whether we are blue-check-marked celebrities or unwashed plebs. Lebesgue lived in Riemann's vertiginous city; you and I live in the strange tiered landscape that Lebesgue redefined. ∎

ETERNITY XIX.

Maria Agnesi holds a glowing orb that looks like
a hairy orb, thanks to the artist's limitations.

XIX.

A GREAT WORK OF SYNTHESIS

I n every branch of mathematics, there is a rule so deep and central that it becomes known as "the fundamental theorem." Oliver Knill, a research mathematician, once compiled more than 150 of these one-line constitutions, from geometry ($a^2 + b^2 = c^2$) to arithmetic (every number has a unique prime factorization) to *Fight Club* ("the fundamental theorem of Fight Club is: you do not talk about Fight Club"). (I'm kidding; Knill, of course, obeys the theorem by not including it.) Anyway, in the whole firmament of fundaments, the greatest theorem of all belongs to— you guessed it!—trigonometry.

No, I jest. Calculus is the head-and-shoulders winner here. And the first mathematician to give the FT of C its proper due was Maria Gaetana Agnesi.

Born in 1718, she was by 1727 fluent in French, Greek, Latin, and Hebrew, not to mention her native Tuscan dialect. She had also won local fame for a speech—which she didn't write but did translate into Latin, memorize, and recite—defending a woman's right to pursue an education, a speech whose best argument, I suspect, was the identity of the person speaking it.

Her father relished the success. A new-money merchant, he saw his daughter's brain as the family's greatest asset, their ticket to noble status.

Soon, Agnesi—the eldest of 21 children—became the featured attraction at dinner parties called *conversazioni*. The literal translation is "conversations," but a more precise rendering is "nerd soirees." Between musical interludes, guests were invited to debate the now-teenaged Agnesi on matters of science and philosophy. She'd open with off-the-cuff orations about Newtonian optics or tidal movements, then roll with it when folks challenged her on metaphysics or mathematical curves. Then, to end the evening: sorbet for everyone! I can't decide if the whole thing sounds as dull as Latin class, or as fun as an ice-cream party in Latin class.

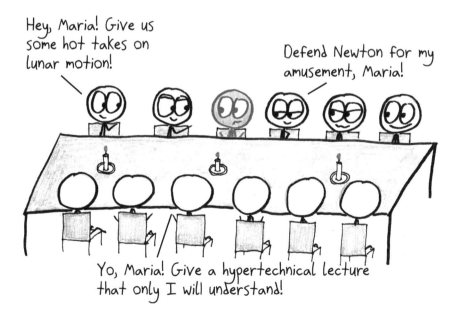

Evidently, Agnesi had mixed feelings herself. Learning, debating, and scientific chitchat? She was all for it. Showmanship, competitiveness, and clawing for social status? Not so much. By age 20, she negotiated with her father to ease up on the *conversazioni*. Instead, she spent her time volunteering at hospitals, teaching illiterate women to read, and helping the poor and infirm. You know, rebellious daughter stuff.

At the age of 30, she published her only book: *Instituzioni Analitiche ad Uso della Gioventù Italiana*, or "Analytical Institutions for the Use of Italian Youth." Originally conceived as a way to teach math to her little brothers, it developed into much more: a way to teach math to *everyone's* little brothers. "She came to believe," says historian Massimo Mazzotti, "she could work on a much more ambitious project: an introduction to calculus that would guide the beginner from the rudiments of algebra to the new differential and integral techniques. This would be a great work of synthesis..."

Indeed: it was to become the most complete, accessible, well-organized book yet written on calculus, and the first work to unify the derivative and the integral in a single volume. That comprehensive sweep allowed Agnesi to place fresh emphasis on a little old fact. What you might call a *fundamental* fact.

So let's get to the FT of C.

For mathematicians, "inverse processes" are actions that undo each other, counteracting opposites. Think of adding 5 and subtracting 5. One brings you from A to B, the other right back from B to A.

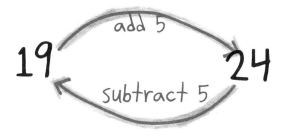

The same goes for tripling and dividing by 3. Pick any number; triple it; then divide by 3. Let me guess: you got back your original number! I must be psychic, right?

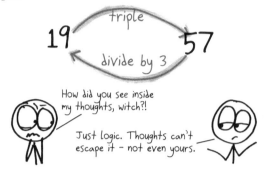

Mathematics is full of counteracting pairs like this. Squaring turns 3 into 9; the square root turns 9 back into 3. The exponential turns 2 into 100; the logarithm turns 100 right back into 2. A year of lessons turns the ignorant brain into a knowledgeable one; summer restores the brain to its original state.

The fundamental theorem of calculus is the plain and startling fact that *the derivative and the integral are opposites*. I don't mean this in some casual, rhetorical way. It's not like saying "Hermione and Ron are opposites" because one is coolheaded and the other hotheaded, one female and the other male, one brilliant and the other Ron. No, I'm using "opposites" in the precise mathematical sense.

I teach this to my students via a physical example. Say we've got a *position* function: we know exactly where our car has been at every moment for the last few hours.

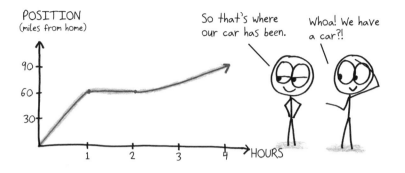

From this information, can we determine the car's *velocity*? Sure! Just look at the slope of the graph. This is called "differentiating" or "taking a derivative."

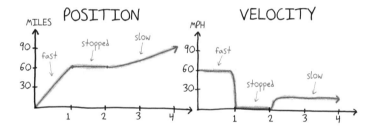

Now, wipe your mental blackboard clean. Imagine instead that we begin with a *velocity* function: we know exactly how fast our car has been going at every moment for the last few hours.

From this information, can we determine how the car's *position* has changed—i.e., how far it has moved? Sure! The distance traveled is simply the area under the curve. This process is called *integrating*, or *taking an integral*.

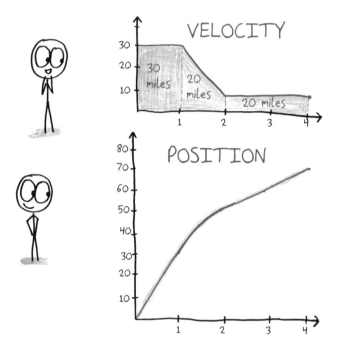

Thus, the derivative and the integral—finding the slope of the curve, and finding the area under it—are opposites. The former extracts a moment from the stream of time; the latter rebuilds the stream from a collection of drops.

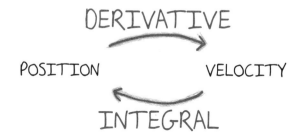

But that's my explanation, not Agnesi's. An 18th-century gal, she didn't go in for automobiles. In fact, in her book, she refused to include physical applications of any kind.

It's not that she was an indifferent or careless teacher. Once, after her father coerced her into giving a daunting technical lecture at a dinner party, she apologized to a guest, saying she "did not like to speak publicly of such things, where for every one that was amused, twenty were bored to death." Nor did she dislike physics in its own right; heck, she was the town expert. So why the ban on physical examples, which help to make calculus concrete and meaningful? Did her brothers never ask for "real-world applications" or wonder, *When are we going to use this?*

Maybe they did. But for Agnesi, mathematics wasn't about practicality. It was a sacred project, a pathway to God. Pure logical thought gave humanity its closest experience of divine cognition, of eternal truth. For someone as devout as Agnesi, that meant everything. Why sully the holy with the earthly, the geometric with the physical?

Agnesi's purified approach birthed an enduring work. "The notation is so well chosen and modern," writes math historian Joaquin Navarro, that "not a single comma needs moving for it to be intelligible to a modern audience." To appreciate Agnesi's perspective on the FT of C, consider an integral as a sum of countless tiny rectangles, nestled under the curve of a graph.

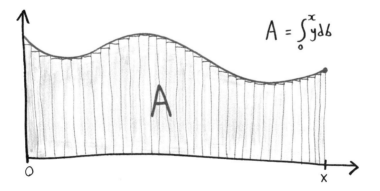

$$A = \int_0^x y \, db$$

The derivative measures a change to this total area—in other words, the size of the last rectangle to join the skyline.

But, in proportion to the infinitesimal dx, this rectangle's size is just the height of the curve.

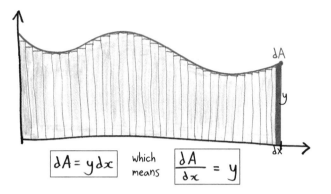

This means that if you (1) start with a curve, (2) integrate under it, and (3) differentiate, then you'll wind up right back where you began. Although seemingly unlike our car-driven discussion before—the two are sometimes distinguished as "the first fundamental theorem" and the "second fundamental theorem"—both roads lead to the same destination. Once again, integrals and derivatives reveal themselves to be like poison and antidote, or pencil and eraser.

According to the FT of C, all of calculus is a giant yin-yang symbol.

Agnesi understood the unity of opposites better than anyone. Just look at the identities she embodied: mathematician and mystic, Catholic traditionalist and proto-feminist, disciple of science and of religion alike. She even bridged the starkest opposition of all, the bitter Newton/Leibniz feud, which was still radioactive when she set out to write. As no one else had, Agnesi managed to unify the Englishman's "fluxions" with the German's "differences," achieving such a perfect fusion that one Cambridge mathematics professor learned Italian expressly so that he could translate her masterwork into English.

She didn't view any of these as contradictions. As Mazzotti writes, "The very categories of 'science' and 'religion' as referring to two incompatible sets of practices would be meaningless to Agnesi." It's our own epoch that opposes reason and faith. Agnesi knew better.

In 1801, when that eager Cambridge professor translated her book, he mistook the word *versiera* (a sailing term for "sheet") for its homonym, an abbreviation of *avversiera* ("she-devil"). Thus, a certain mathematical curve became known to English speakers as the "Witch of Agnesi." It's an enduring testament to either Maria's brilliance or the dangers of amateur translation.

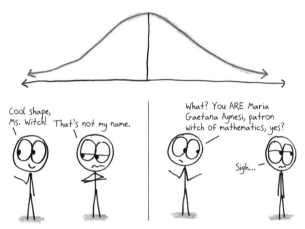

Today, the fundamental theorem of calculus is perhaps the most powerful and ubiquitous shortcut in mathematics. With it, an integral—that delicate sum of infinite pieces, each infinitesimally small—becomes a simple antiderivative. We can forget the intricate skylines of Riemann, the subtle rearrangements of Lebesgue, the geometric maneuvers of Eudoxus and Liu Hui. Instead, just take a backward derivative. It's as if, after years of systematically dismantling the door every time we entered a house, we finally learned about keys.

But that's not why Agnesi celebrated it. "Calculus was, to her, a way to sharpen the mind so it can appreciate God," writes Mazzotti. "She believed in a clear-eyed spirituality, not baroque piety or fantasy-infused superstition." It's through those clear eyes that we get our best glimpse of calculus, a discipline both practical and intrinsically beautiful.

We are all, in that sense, Italian youth. ■

ETERNITY XX.
Another raucous party in the integrand.

XX.

WHAT HAPPENS UNDER THE INTEGRAL SIGN STAYS UNDER THE INTEGRAL SIGN

R ichard Feynman hated math class. The problem was that the teacher always gave you, right alongside the question to solve, the method by which to solve it. Where's the adventure in that? The endeavor felt plodding and bloodless—a government of the dullards, by the dullards, and for the dullards.

Math *club*, on the other hand, he loved. It was a playground, a school of trickster witchcraft and improvisational wizardry. The problems required only algebra (no calculus), but each possessed a devious twist. If you tried to apply a standard method, you'd run out of time. Instead, you had to discover a simplifying shortcut. For example…

QUESTION: You're rowing upstream at $4\frac{1}{3}$ mph, against a current of 3 mph. At 12 p.m., you drop your hat overboard, to be carried along in the current. At 12:45 p.m., you reverse direction. What time do you reunite with your hat?

Sure, you can churn through the arithmetic. But it's swifter to shift your perspective, to make the river current your frame of reference. To become the hat.

ANSWER: The boat travels away from the hat at $4\frac{1}{3}$ mph, and it travels back at the same speed. Thus, the return journey takes the same 45 minutes as the outward one, resulting in a 1:30 p.m. rendezvous.

Derivatives are a bit like Feynman's math class. In any textbook worth the paper it's printed on—and even in several that aren't—you'll find a complete and definitive list of differentiation formulas. Apply those laws, and you can't go wrong.

What about integrals? Per the fundamental theorem of calculus, they are antiderivatives—derivatives taken backward. The derivative of x^2 is $2x$; the integral of $2x$ is x^2. But as you know if you've ever tried to unbake a cake, unshatter a vase, or—what is truly impossible—cancel a magazine subscription, undoing is trickier than doing. Integration, by the same token, is a buffet of spicy exceptions. It's the calculus equivalent of math club.

"No matter how extensive the integral table, it is a fairly uncommon occurrence to find in the table the exact integral desired," says the refreshingly self-effacing *Standard Mathematical Tables*. For example, check out these two: $\int \frac{1}{1+x^2}dx$ and $\int \frac{1}{1+x^3}dx$. Don't sweat the particulars: simply observe that they are sibling questions, and thus ought to yield sibling answers. At least, answers of sibling complexity. So since the former is in my table as **arctan(x)**, the latter should be...

Um...

Checking my notes...

Okay, checking the internet...

Ah, I should have guessed...

It's $\frac{1}{6}\left(-\log(x^2 - x + 1) + 2\log(x + 1) + 2\sqrt{3}\arctan\left(\frac{2x - 1}{\sqrt{3}}\right)\right)$.

My, how the *Standard Mathematical Tables* have turned.

If differentiation is a government building, with bright bureaucratic lights and neatly labeled conference rooms, then integration is a haunted funhouse full of strange mirrors, hidden staircases, and sudden trapdoors. There are no airtight rules to get you safely through—just a scattered collection of diverse tools.

The mathematician Augustus De Morgan, with a poetic flourish, put it like this:

> *Common integration is only the memory of differentiation. The different artifices by which integration is effected are changes, not from the known to the unknown, but from forms in which memory will not serve us to those in which it will.*

At first glance, a novice wouldn't know what to do with $\int \frac{4x^3 + 4x}{x^4 + 2x^2 + 5}dx$. But perform a "change of variables"—a bread-and-butter integration technique—and this vexing riddle becomes the rather unvexing $\int \frac{du}{u}$, which is found in any table of common integrals. Nothing has really changed: just the language employed, the variable's name.

The solution is to shift your frame of reference. To become the hat.

Teaching himself integration in the back corner of his high school physics classroom, Feynman never learned some of the standard techniques. Instead, he gathered tools from off the beaten path: nifty yet little-taught maneuvers like "differentiating under the integral sign."

"I used that one damn tool again and again," Feynman later wrote, after winning a Nobel Prize in physics.

At MIT and Princeton, Feynman's peers would come to him with integrals they couldn't unravel. Feynman would solve them, leaning often on that one powerful trick. "I got a great reputation for doing integrals," he wrote, "only because my box of tools was different from everybody else's." With derivatives, everyone dances the same choreography, but integrals lend themselves to personal style.

During World War II, Feynman joined the flow of scientists toward Los Alamos National Laboratory. He bounced from division to division, learning the ropes, feeling altogether useless. Then, one day, a researcher showed him an integral that had stumped the team for three months. "Why don't you do it by differentiating under the integral sign?" asked Feynman. The problem yielded in half an hour.

Never having learned the technique myself, I went a-Googling. My inquiries led to Math 55 at Harvard: "probably the most difficult undergraduate math class in the country," per Wikipedia. Course alumni include Fields medalists (e.g., Manjul Bhargava), Harvard faculty (e.g., Lisa Randall), and Bill Gates (e.g., Bill Gates). "It's definitely a cult," Raymond Pierrehumbert, a former pupil (and current Oxford professor) told the *Harvard Crimson* in 2006. "I view it as more of an ordeal than a course," he said. Inna Zakharevich, now a professor at Cornell, has fonder memories. "It made me do my favorite kind of thinking," she says, "taking a basic thing that I thought I knew, and thinking really, really, really hard about it."

In 2002, the 18-year-old Zakharevich had just read Feynman's memoir. "I didn't know what differentiation under the integral was. I asked my dad about it, and we discussed the general approach." Then, one October day, Math 55 professor Noam Elkies showed the class a formula: $n! = \int_0^\infty x^n e^{-x} dx$.

In mathematics, the ! symbol does not express enthusiasm. It stands for the operation "factorial," which means "multiply all the counting numbers up to that number."

$$3! = 3 \times 2 \times 1$$

$$5! = 5 \times 4 \times 3 \times 2 \times 1$$

$$100! = 100 \times 99 \times 98 \times$$
$$\dots \text{ and so on...} \times 2 \times 1$$

Very cool. Very emphatic-looking. But, as defined, very limited: it makes sense only for whole numbers.

$$7.26! = \dots ?$$

Um... what?

In the 1700s, Leonhard Euler came up with a new way to define factorials: the integral Elkies showed Math 55. It promised to extend the factorial concept to all numbers, allowing you to compute $\pi!$ or $1.8732!$ or $\sqrt{2}!$ to your heart's content.

$$3! = \int_0^\infty x^3 e^{-x} dx$$

Ahh, what a glorious extension!

$$11! = \int_0^\infty x^{11} e^{-x} dx$$

$$7.26! = \int_0^\infty x^{7.26} e^{-x} dx$$

Just one problem: Are we sure the new definition matches the old one? How do we know they're equal for numbers like 3 and 11?

Zakharevich watched Elkies deploy the standard demonstration of the equality: a repeated, slogging application of *integration by parts*. It's a ubiquitous and, in this case, rather unwieldy technique. "I was frustrated," Zakharevich remembers, "because it was such an ugly proof."

Ever the dutiful pupil, she regurgitated the gnarly algebra on that day's quiz. But on the back, she wrote up an alternate proof, drawing on Feynman's favorite technique. "I really wanted Elkies to know it for the future," she explains. In the proof—which I reproduce below, more for decoration than anything—Zakharevich introduces a new parameter, takes a derivative with respect to it, and then lets it shuffle back into the shadows. It's like a roadside helper who changes your flat tire and then vanishes before you can even say thanks.

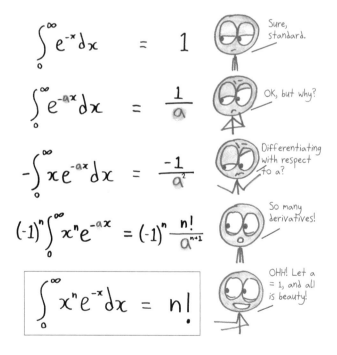

Elkies loved it. Glowing with teacherly pride, he posted it online, where 16 years later I stumbled across it.

"Applying it," she admits, "is really an art, rather than a science."

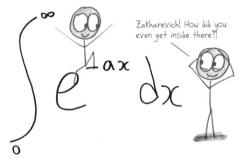

I'm sure Feynman would approve. It's the defeat of math class and the triumph of math club—of the trickster approach that, for him, encompassed everything in life. Take his later service on school boards, as recounted by biographer James Gleick:

> *He proposed that first-graders learn to add and subtract more or less the way he worked out complicated integrals—free to select any method that seems suitable for the problem at hand. A modern-sounding notion was,* The answer isn't what matters, so long as you use the right method. *To Feynman no educational philosophy could have been more wrong. The answer is all that does matter, he said... Better to have a jumbled bag of tricks than any one orthodox method.*

Feynman loved to show off his own bag of tricks. Once, he dared his Los Alamos colleagues to state any problem in 10 seconds, promising to compute a solution in less than a minute, to within 10% accuracy. His friend Paul Olum punctured his pride by asking for the tangent of a googol, which required figuring $\frac{1}{\pi}$ to one hundred digits: too much, even for a future Nobel laureate.

Another time, Feynman boasted that whatever you could solve by the traditional method of "contour integration," he could solve by other techniques. He defeated several challengers, and failed only when Olum, the perfect nemesis, posed "this tre*men*dous damn integral... He had unwrapped it so that it was *only* possible by contour integration!" Feynman recalls. "He was always deflating me like that." That's the joy and frustration of integrals: no one, except perhaps Paul Olum, ever has all the tricks. ∎

ETERNITY XXI.
Einstein makes a cosmic blunder.

XXI.

DISCARDING EXISTENCE WITH A FLICK OF HIS PEN

By 1917, Albert Einstein had already made quite a name for himself: specifically, the name "Einstein." He had calculated the size of atoms, established the equivalence of matter and energy, launched quantum physics, and cultivated a hairstyle best dubbed "frizz nova." An impressive résumé, but his proudest achievement, far and away, was his theory of general relativity. It was an equation of singular elegance. It was a mushroom trip for the cosmos. It was a roundhouse kick to the face of Newtonian mechanics. It was a reality so strange that the *New York Times* worried it might call into question "even the multiplication table." And it all built upon a simple insight: The universe is not a box in which the stars and planets reside. It bends, it warps, in the presence of matter.

Time for a thought experiment. Imagine that I'm lazing on a stump, watching a beam of light blaze past, at its invariable speed of 300 million meters per second. Meanwhile, you chase that beam at a tremendous speed, moving past me at, say, 200 million meters per second.

Does the light recede faster from me or from you?

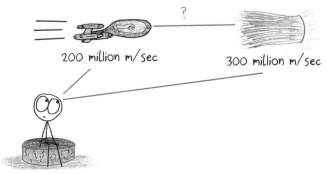

200 million m/sec

300 million m/sec

Trick question! The speed of light is a universal constant. It's always 300 million meters per second, not changing for nobody, not nohow—not even superspeedy you. What changes, instead, is something softer and more pliable: the fabric of space and time. From my vantage here on the stump, it takes three seconds for light to open up a 300-million-meter lead on you. From your vantage on the *Starship Enterprise*, that event takes only a single second. Thus, my pocket watch is advancing at thrice the speed of yours.

Motion reshapes time.

Hallucinating yet? Then you're ready for the next step: matter reshapes space, too. The sun, for example, doesn't just sit there like a bowling ball in a box. It sits there like a bowling ball on a mattress, weighing on the fabric, warping the surrounding region of space-time. Thus, when a planet orbits the sun, or when an apple falls to earth, it is not in the throes of some unexplained Newtonian attraction. It's just following the path of least resistance through a curving four-dimensional landscape.

"Matter tells space-time how to curve," said physicist John Wheeler, "and curved space tells matter how to move."

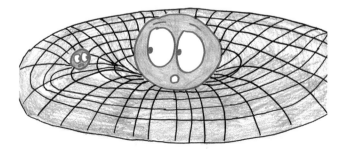

All of this was crystallized in November 1915, in the form of the Einstein field equation. "The equation fits into half a line," writes physicist Carlo Rovelli. "But within this equation there is a teeming universe." It predicted that light would bend around heavy objects, that time's flow would dilate from valley to mountaintop, that gravitational waves could propagate through the universe, that large stars could collapse into singularities (later dubbed "black holes"). "A phantasmagorical succession of predictions that resemble the delirious ravings of a madman," writes Rovelli, "but have all turned out to be true."

Still, even as new predictions issued from this equation like Patronuses from a magic wand, Einstein remained unsatisfied. Sure, general relativity could describe orbiting planets and bending photons. But those were finite, bounded systems. Mere pieces of the cosmos. "It was a burning question," Einstein wrote to a colleague, "whether the relativity concept can be followed through to the finish or whether it leads to contradictions." He now sought the grandest prize of all, the biggest teddy bear at this or any carnival.

Could general relativity model the entire universe?

It was a question in the spirit of integration, the leap from "lots of little somethings" to "a whole big everything." In fact, it literally involved integration; although his famous 1917 paper took a different approach, by 1918 Einstein had discovered that he was, in effect, taking an integral. He preferred this framing. "The new formulation has this great advantage," he wrote, "that the quantity…appears in the fundamental equations as a constant of integration."

What quantity? We'll get there. First: What *is* a constant of integration?

If you ask a calculus student, it's the annoying +C at the end of every indefinite integral. It's a notational flourish, irrelevant to the integral you're computing, but which, per some obscure regulation, you must never forget, lest your petty bureaucrat of a teacher dock points.

Where does the constant come from? Well, as we've discussed, integration and differentiation are inverse processes. To take an integral, we look at a function and ask, What is this the derivative of?

Consider a runner moving at the steady speed of 7 miles per hour. The velocity graph looks like this:

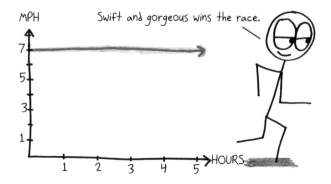

What about the integral—that is to say, the position graph? Well, here's one possibility:

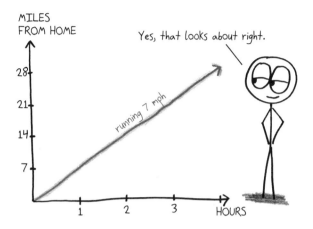

But this assumes that, at noon, the runner set out from home. Really, we don't know where the run began. Perhaps it was a mile from home, or two, or seven. Or three and a half, but on the opposite side of home, thereby passing home at 12:30 p.m.

There are infinite possible position functions, each identical, except for a certain fixed distance added or subtracted. It could be $7x$, or $7x + 1$, or $7x + 2$, or $7x + 3$...

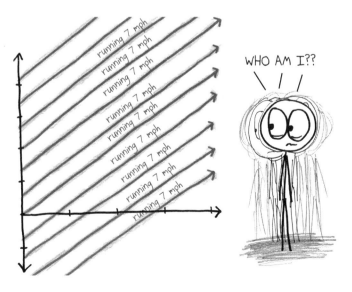

Rather than list the infinite possibilities, which could delay dinner, we summarize the whole family with the simple formula $7x + C$. The C is a constant of integration, a shorthand for "any number whatsoever." In one stroke, it transforms a single curve into an infinite family. The same concision that makes it easy to forget is the source, too, of its power and profundity.

Now, Albert Einstein didn't forget the constant. I mean, c'mon—we're talking about one of the greatest scientists ever to forgo a comb.

No, he committed an error far more deliberate, and far more spectacular.

"I shall conduct the reader over the road that I have myself travelled," Einstein writes in his 1917 paper, "rather a rough and winding road." Indeed, he navigates something like a maze of one-way streets, thwarted at every mathematical turn. His first attempt to describe the whole universe contradicts known facts. His second would require specifying a "correct" frame of reference, against the whole spirit of "relativity." And a third path, suggested by a colleague, "holds out no hope of solving the problem but amounts to giving it up." His celebrated equation just isn't giving him enough flexibility.

In the end, Einstein is able to salvage his model only by introducing a constant of integration: Λ. That's the Greek letter lambda; Einstein actually used the lowercase λ, perhaps signaling his lack of respect for it. Anyway, lambda is the cosmological constant.

It was a perfectly valid mathematical move, and a necessary one, too: without λ, the model collapsed. It predicted either a shrinking universe (if there was lots of matter around), an expanding universe (if there wasn't very much), or a totally matter-free universe (which stayed the same size, but in a sad, empty sort of way). Only a specific fine-tuned value for λ allowed Einstein to describe the universe he knew: one that contained matter and didn't change sizes.

Still, Einstein felt ambivalent. The whole paper reads a bit like an apology for λ. He saw it as a blemish on the theory, an inelegant complication. Its necessity frustrated him, as if his car's engine would only run properly with a certain hood ornament.

That's how affairs stood for a decade or more. Then, in 1929, big news came down from astronomer Edwin Hubble. In fact, as measured in cubic meters, it was the biggest news ever.

What everyone had called "the universe" was not the universe. It was just our galaxy, the Milky Way. The blurry spiral nebulae in the night sky were in fact other galaxies, similar in scope to ours, yet millions of light-years away, most of them racing even farther into the distance. Thus, not only was the universe vastly bigger than we had imagined it, but it was expanding every moment. The galaxies were separating like raisins in a rising loaf of bread.

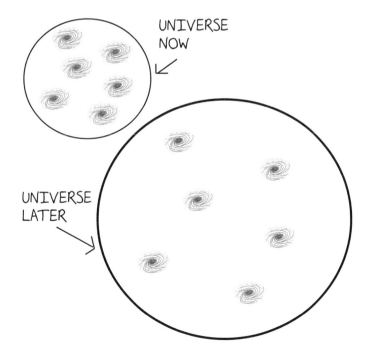

The expanding universe meant that—although it didn't necessarily have to—λ could now equal zero. That was enough for Einstein. Without a moment's hesitation or sentimentality, he ditched λ, calling it "theoretically unsatisfactory," and claiming it was zero. (Possibly relevant: Einstein was really nasty in romantic breakups.) "If Hubble's expansion had been discovered at the time of the creation of the general theory of relativity," he later wrote, "the cosmologic member would never have been introduced." According to his friend George Gamow, Einstein confided "that the introduction of the cosmological term was the biggest blunder he ever made in his life."

Some argue that he blamed λ for his failure to predict the universe's expansion, which would have been a crown jewel for general relativity. But there's little evidence he felt that way. He ventured into cosmology with the narrow aim of proving general relativity could build a consistent model, and never lamented that "lost prediction." Rather, his anti-λ grudge seems to flow from an aesthetic preference that constants of integration should equal zero, rather like those people who insist that children ought to be neither seen nor heard.

Whatever the reason for his "biggest blunder" comment, the true blunder was the comment itself.

In 1998, it emerged that the universe isn't just expanding. The expansion is accelerating. In a single stroke, this revived the cosmological constant from a half century of dormancy. It even returned as a capital letter.

Now, it seemed Λ wasn't zero after all: it captured the existence of "dark energy," a peculiar presence that inhabits empty space, pushing against gravity. By our current understanding, it comprises some 68% of the contents of the cosmos.

Einstein's constant of integration wasn't a blunder to be swept aside. It was, quite literally, two-thirds of the universe.

No one ever claimed Albert Einstein was a flawless mathematician, least of all Albert Einstein. "Do not worry about your difficulties in Mathematics," he wrote to a 12-year-old pen pal. "I can assure you mine are still greater." A book called *Einstein's Mistakes*—please, no one write a book called *Orlin's Mistakes*—claims that perhaps 20% of his papers contain substantive errors. Mr. Frizz Nova took it all in stride: "He who has never made a mistake," he quipped, "has never tried something new."

That's how it is with constants of integration. They're easy to neglect, hard to interpret, and sometimes they really are zero. Other times, they encode crucial information. The beginner may forget a constant of integration; the expert, by contrast, remembers it, then goes back and erases it, insisting that it had to be zero all along.

I don't know about you, by for my part, Einstein's tale makes me grateful to occupy this curving, expanding acid trip of a universe, where even the constants tell stories of change. ∎

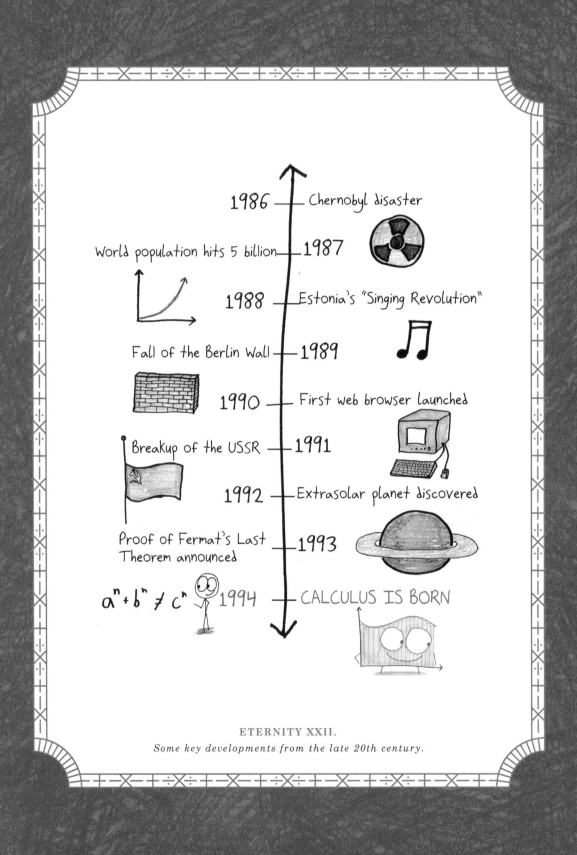

1986 — Chernobyl disaster

World population hits 5 billion — 1987

1988 — Estonia's "Singing Revolution"

Fall of the Berlin Wall — 1989

1990 — First web browser launched

Breakup of the USSR — 1991

1992 — Extrasolar planet discovered

Proof of Fermat's Last Theorem announced — 1993

$a^n + b^n \neq c^n$ 1994 — CALCULUS IS BORN

ETERNITY XXII.
Some key developments from the late 20th century.

XXII.

1994, THE YEAR
CALCULUS WAS BORN

In February of the 1994th year of the Common Era, the medical journal *Diabetes Care* published an article by researcher Mary Tai. Its title: "A Mathematical Model for the Determination of Total Area Under Glucose Tolerance and Other Metabolic Curves."

Sensationalist clickbait, I know, but bear with me.

Whenever you eat food, sugar enters your bloodstream. Your body can make glucose out of anything, even spinach or steak, which is why the "Orlin Diet" skips the middleman and prescribes only cinnamon rolls. Whatever the meal, your blood-sugar level rises and then, over time, returns to normal. Key health questions: How high does it rise? How fast does it fall? And, most of all, what trajectory does it follow?

The "glycemic response" is not just a peak or a duration; it's a whole story, the aggregate of infinite tiny moments. What doctors want to know is the area under the curve.

Glucose Level Over Time

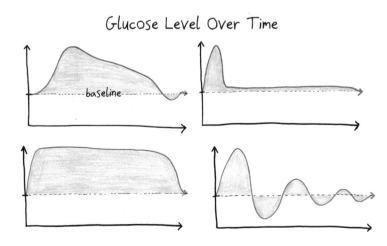

Alas, they can't just invoke the fundamental theorem of calculus. That's for curves defined by tidy formulas, not ones born from a game of connect-the-dots with empirical data. For such messy realities, you need approximate methods.

That's where Tai's paper comes in. "In Tai's Model," it explains, "the total area under a curve is computed by dividing the area...into small segments (rectangles and triangles) whose areas can be accurately calculated from their respective geometrical formulas."

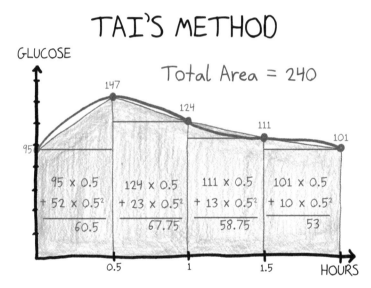

Tai writes that "other formulas tend to under- or overestimate the total area under a metabolic curve by a large margin." Her method, by contrast, appears accurate to within 0.4%. It's clever geometric stuff, except for one tiny criticism.

This is Calc 101.

Mathematicians have known for centuries that, when it comes to practical approximations, there are better methods than Riemann's skyline of rectangles. In particular, you can identify a series of points along your curve, then connect the dots with straight lines. This creates a procession of long, skinny trapezoids.

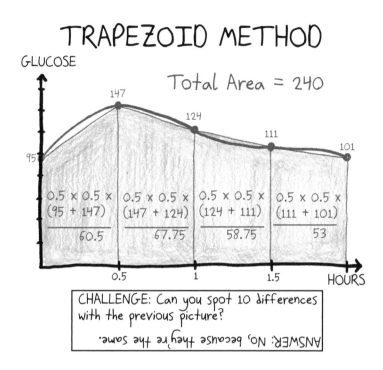

TRAPEZOID METHOD

GLUCOSE

Total Area = 240

147
124
111
101
95

0.5 × 0.5 × (95 + 147)	0.5 × 0.5 × (147 + 124)	0.5 × 0.5 × (124 + 111)	0.5 × 0.5 × (111 + 101)
60.5	67.75	58.75	53

0.5   1   1.5   HOURS

CHALLENGE: Can you spot 10 differences with the previous picture?

ANSWER: No, because they're the same.

Forget 1994. This method wasn't new in 1694, or in 94 BCE. Ancient Babylonians employed it to calculate the distance traveled by the planet Jupiter. Tai had written, and a referee had approved, and *Diabetes Care* had published, a piece of work millennia old, one that a gung-ho undergraduate could do for homework. All as if it were novel.

Mathematicians had a field day.

Event #1: Headshaking. "Tai proposed a simple, well-known formula exaggerately [*sic*] as her own mathematical model," wrote one critic in a letter to *Diabetes Care*, "and presented it in a circumstantial and faulty way."

Event #2: Sneering. "Extreme ignorance of mathematics," one online commenter remarked. "This is hilarious," wrote several others.

Event #3: Reconciliation. "The lesson here is that calculating areas under the curve is deceptively difficult," wrote a diabetes researcher whose earlier paper Tai had criticized (based on an incorrect understanding, it turns out). The letter ended on a conciliatory note: "I fear I may be responsible for contributing to the confusion."

Event #4: Teachable Moments. Two mathematicians pushed back on Tai's insistence that her formula was not about trapezoids, but about triangles and rectangles. They even drew a picture for her: "As is evident in the following figure...the small triangle and the contiguous rectangle form a trapezoid."

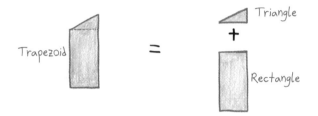

Event #5: Soul-Searching. "As a smug physicist myself," wrote one commenter, in response to a fun-poking blog post, "I did find this funny, but I can't help thinking that this post makes us look worse than them....I'm sure you can find plenty of physicists saying spectacularly naïve things about medicine or economics."

For what it's worth, mathematical researchers have been known to reinvent the wheel, too. During his graduate studies, the legendary Alexander Grothendieck rebuilt the Lebesgue integral for himself, not realizing he was replicating old work.

As Tai tells it, she wasn't trying to glorify her method. "I never thought of publishing the model as a great discovery or accomplishment," she wrote. But colleagues "began using it and…because the investigators were unable to cite an unpublished work, I submitted it for publication at their requests." She was just trying to share her thinking, to facilitate further inquiry.

Alas: In academia, publishing isn't just about sharing information. It's more than a way to say *Here's a thing I know*. It's also a scoreboard, a declaration of *Hey, here's a thing I know because I discovered it, so please esteem me highly. Thank you and good night.*

This publication system has its flaws. "Our basic and most important unit of discourse is a research article," writes mathematician Izabella Łaba. "This is a fairly large unit: effectively, we are required to have a new, interesting and significant research result before we are allowed to contribute anything at all."

Łaba likens this to an economy where the smallest denomination is the $20 bill. In such a world, how can anybody run a bake sale? Either you've got to coax folks into buying $20 worth of muffins, or you've got to give them away for free. Tai chose to charge $20, but neither option is much good. "We should have smaller bills in circulation," writes Łaba. "It should be possible to make smaller contributions—on a scale of, say, a substantive blog comment."

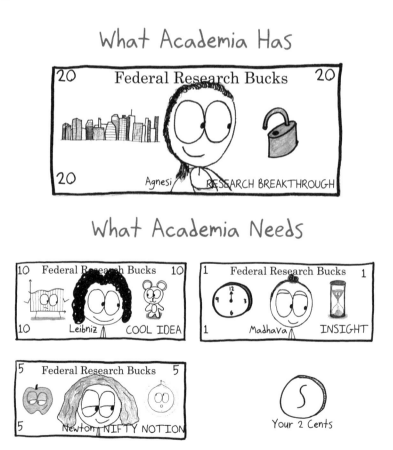

Integrals aren't just for mathematicians. Hydrologists use them to estimate the flow of contaminants through groundwater; bioengineers, to test out theories of lung mechanics; economists, to analyze a society's income distribution, its deviation from a state of perfect equality. Integrals belong to diabetes researchers, to mechanics, to nutjob Russian novelists, to anyone and everyone who seeks the area under a curve, the infinite sum of infinitesimal pieces. The integral is a hammer in a world full of nails, and it doesn't just belong to hammer makers.

But calculus teachers, like your sheepish and regretful author, can slip up. We emphasize the antiderivative approach—which works only in theoretical settings, where we possess an explicit formula for the curve. That privileges philosophy over practice, abstraction over empirics.

It's also outdated. "Numerical analysis has grown into one of the largest branches of mathematics," writes professor Lloyd N. Trefethen. "Most of the algorithms that make this possible were invented since 1950." The trapezoid rule, alas, is not among them. But calculus, for all its antiquity, remains a growing field, even in the years since 1994. ∎

ETERNITY XXIII.

A moral philosopher's laboratory.

XXIII.

IF PAINS MUST COME

"**N**ature has placed mankind," declared Jeremy Bentham in 1780, "under the governance of two sovereign masters." Eschewing the obvious candidates (pecan pie and afternoon naps), he named the twin sovereigns as *pleasure* and *pain*. I suppose it makes sense. From there it's only a short step to embrace Bentham's conclusion: that we ought to limit pain to a minimum and spread pleasure to a maximum.

Thus was born *utilitarianism*: a philosophy that, like every philosophy, is more headache-inducing than it first appears.

Utilitarianism asks us to seek the greatest good for the greatest number. Better to give 11 back rubs than 10. Better to slap zero faces than one. Easy enough. But what if a pleasure and a pain stand opposed? Imagine, for a highly plausible example, that we're somehow saving lives by kicking people in the shins. How should we navigate the resulting trade-offs? Surely a single life saved is worth 50 shin kicks. Even 500. But what about 50,000, or 5 million? What if we need to kick every shin on Earth to save a single life? If the shin pain lasts a minute, then that's nearly two hundred lifetimes of human hurt, spread across the globe, all to rescue one life. Is it better, then, to sacrifice the one and spare the many?

Should one man die for mankind's shins?

Utilitarianism reduces ethics to a kind of mathematics, what philosophers call "felicific calculus." To judge a prospective action, we need to quantify the pleasure and the pain that it will cause, so that they can be weighed against one another. Bentham helpfully outlined the relevant considerations:

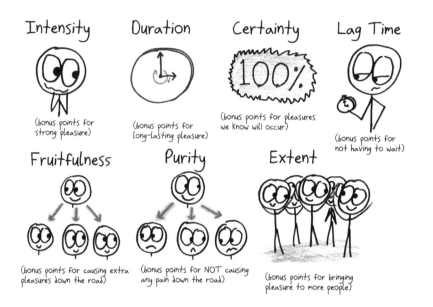

He even composed a little ditty for legislators to memorize, a mnemonic to help keep these criteria in mind as they debated laws:

Intense, long, certain, speedy, fruitful, pure—
Such marks in pleasures and in pains endure.
Such pleasures seek if private be thy end:
If it be public, wide let them extend
Such pains avoid, whichever be thy view:
If pains must come, let them extend to few.

Hey, I love a good poem. I even love a bad poem. But a part of me wishes Bentham would write a little more in the style of—I never thought I'd say this—an algebra textbook. "Deal with the soul," Emily Dickinson once wrote, "As with Algebra!"

Bentham agrees, yet refuses to act like a proper algebra teacher. Where are the numbered exercises, the worked examples, the tidy theorems in shaded boxes? It took a full century before an economist named William Stanley Jevons took a stab at it, seeking to build a felicific calculus out of *actual* calculus.

As a first step, he declared that the *y*-axis will depict the intensity of an emotion:

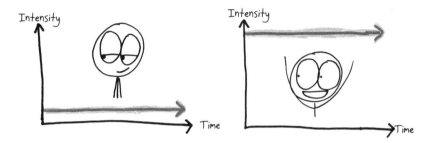

The *x*-axis, meanwhile, will capture the duration of an emotion.

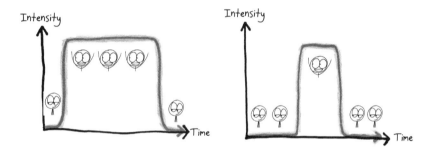

Let's say you are listening to Outkast's "So Fresh, So Clean," a track that lasts exactly four minutes and sustains a constant pleasure level of "irresistibly cool." Computing the joy that this experience confers is thus a simple act of multiplication, like finding the area of a rectangle:

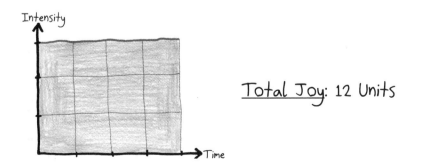

"But if the intensity...," writes Jevons, "varies as some function of the time"—which, with groups less perfect than Outkast, is often the case—then "the quantity of feeling is got by infinitesimal summation or integration."

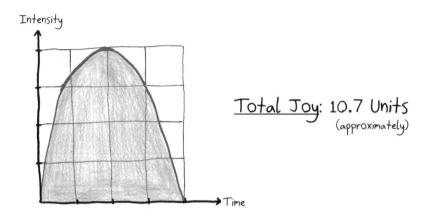

In Jevons's model, a stellar two-minute back rub can somehow "equal" a pretty-good five-minute one. Two hours of kinda needing to pee might be "equal" in some sense to 30 minutes of desperately needing to go.

Robert Frost wrote, in one poem's title, that "happiness makes up in height for what it lacks in length." Jevons makes that relation explicit, mathematical.

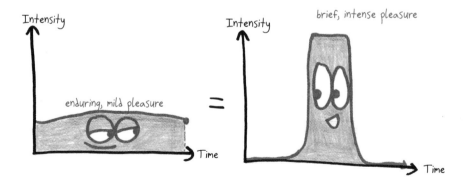

Jevons also asserts that pain and pleasure can cancel out, neutralizing one another. Although some utilitarians disagree—arguing that "hedons" (units of pleasure) and "dolors" (units of pain) are as incomparable as apple juice and orange sweaters—Jevons declares that pleasure and pain are simply "opposed as positive and negative quantities."

If Bentham reduced morality to mathematics, then Jevons aimed to go another step, reducing it to simple questions of measurement. Ethics becomes an exercise in data gathering. If Jevons's scheme works, then doing the right thing will be as easy as weighing a package or totaling a grocery bill. He promises to rearrange the infinite moments of our lives, each infinitesimally brief, into a single integral, a clear moral structure, an achievement that you might reasonably call the greatest breakthrough in the history of human morality.

Which is a giveaway, I suppose, that it's not going to work.

A century after Jevons, a team of psychologists led by Daniel Kahneman set out to study a particular experience of pain: forcing people to hold their hands immersed in icy water. (Psychology: it's sociology for sociopaths.) One hand was submerged in 57°F water for a minute. At another time, the other hand underwent the same experience, followed by an additional 30 seconds in the water, during which the temperature gradually rose to 59°F.

Later, subjects were asked: Which trial would you rather repeat?

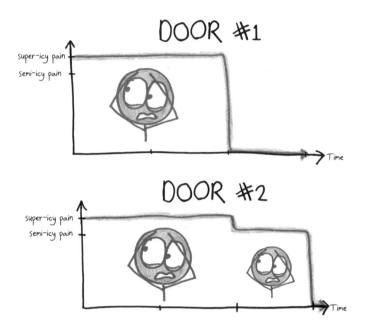

Jevons's theory tells us that nobody should choose the latter trial. It has all the chilly pain of the first, plus a little extra. Unless you're an Arctic mammal, a masochist, or both, bonus icy hand time should not appeal to you.

And yet that's exactly what most subjects chose. Looking back on an experience, people tend to ignore how long it lasted. Instead, they focus on *extremes* and *endings*—the maximum of pain, and the final pain level. Because the second trial hits the same extreme and ends on a slightly less painful note, subjects recall it more fondly.

Emotion, as retained in human memory, is not a Jevons-like integral. It overweighs finales. I'm reminded of Ray Bradbury's insight: "A bright film with a mediocre ending is a mediocre film. Conversely, a medium-good film with a terrific ending is a terrific film." What makes a story happy or sad, cynical or hopeful, tragic or comic? It's the ending, and nothing else. That's why we rush to visit deathbeds, why we dwell on final words, why the last minutes of a lifetime can redefine the eight decades prior.

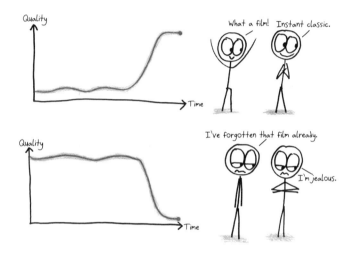

The foundation upon which utilitarianism builds is subjective experience. Human emotion. At times, this foundation seems less like firm bedrock, and more like an active magma flow. It's a serious challenge to the dream of turning morality into mathematics.

Even so, utilitarianism remains a booming and necessary voice in our moral sphere. Sure, we may debate what ranks as "the greatest good" ("better to be Socrates dissatisfied than a fool satisfied," said 19th-century economist John Stuart Mill), or who counts among that "greatest number" ("most human beings are speciesists," warns philosopher Peter Singer), or how to aggregate billions of subjective experiences into a single sum (maybe Tolstoy can help?). We probably reject the specifics of Jevons's moral calculus. But every time we posit a moral calculus of our own—a new model better matched to the complexities of emotional reality—we tread in Jevons's footsteps. Explicit or not, consistent or not, we live our lives by a kind of felicific calculus. ∎

ETERNITY XXIV.

The Claw of Archimedes:
probably apocryphal, definitely awesome.

XXIV.

FIGHTING WITH THE GODS

You know the Romans: a hard-nosed, humorless, our-marble-garbage-will-linger-here-for-millennia kind of people. In the year 212 BCE, their forces came to the Sicilian coast to conquer the stubborn little city of Syracuse. As historian Polybius relates, they were armed to the teeth, in 60 ships "filled with archers, slingers, and javelin-throwers," not to mention four huge boat-mounted siege ladders.

But Syracuse knew the old adage "When in Rome's grip, do as the Romans do." That is, fight like hell. And so, from catapults great and small, the Syracusans launched "immense masses of stone," "large lumps of lead," and a shower of iron darts. Then, great mechanical claws emerged from the city walls and grabbed the Roman ships, which were thus "dashed against steep rocks" and "plunged…to the bottom of the sea." The historian Plutarch narrates: "The Romans, seeing that infinite mischiefs overwhelmed them from no visible means, began to think they were fighting with the gods."

Even worse. They were fighting with Archimedes.

ARCHIMEDES

SCHEMER, DREAMER, ALL-STAR TEAMER

In your office's greatest-mathematician-of-all-time pool, Archimedes is a pretty solid first-round choice. Galileo called him "superhuman." Leibniz raved that he spoiled genius itself, making later thinkers seem prosaic by comparison. "There was more imagination in the head of Archimedes," Voltaire wrote, "than in that of Homer." Admittedly, he never won a Fields Medal, math's most celebrated prize, but in Archimedes's defense, his face is the one on the medallion.

You want a sense of his cleverness? Here: take a cube and cut it into three careful pieces.

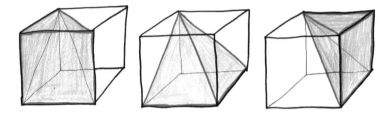

The three figures are identical pyramids, each with a square-shaped bottom and a pinnacle above one corner of the base. Therefore, each must occupy exactly 1/3 of the original cube's volume.

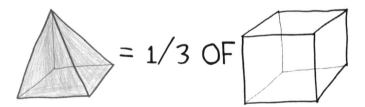

So far, so nifty. But we're just getting started.

Take one of those pyramids and dice it up into infinite slices, each infinitesimally thin. If I've done this right—and, given my clumsiness with ordinary kitchen knives, you may want to double-check my work with this infinite conceptual knife—every cross-section should be a perfect square.

The very bottom square fills the whole base of the cube. The very top one is so tiny that it's just a single point. Between these extremes lie squares of intermediate size.

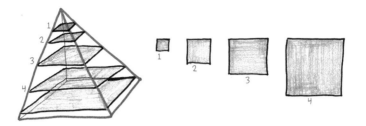

Now, turn up the heat further. Imagine those squares as a stack of infinite cards, each shadow-thin. Rearranging them won't change the volume of the stack, so let's get shuffling. At the moment, our squares share a common corner. But why not nudge them over, so that they share a common *center*? This renovates our funky asymmetric pyramid into a classic Egyptian-style one.

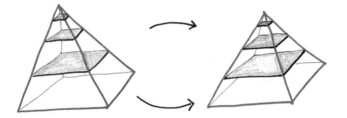

Best of all, the volume doesn't change. It remains 1/3 of the cube.

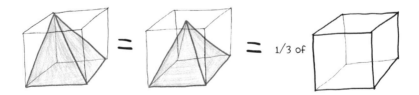

We arrive now at a move so ingenious and handy that, when mathematician Bonaventura Cavalieri rediscovered it 1800 years later, they'd dub it "Cavalieri's principle" in his honor. In reality, it originated with Antiphon (5th century BCE), grew with Eudoxus (4th century BCE—in fact, he first gave the argument I'm giving now), and reached masterful heights with Archimedes (3rd century BCE—we'll get to his singular contributions soon). I'm going to call it, in honor of Roman panic, "the Principle of Infinite Mischiefs."

The idea is simple. In a 3D shape, you don't affect the volume when you swap out cross-sections for others with equal area. For example, we can cash in our squares for rectangles. The now-elongated pyramid still fills 1/3 of the Prism Formerly Known as Cube.

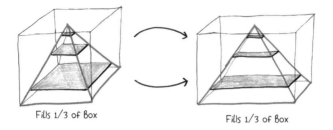

Fills 1/3 of Box   Fills 1/3 of Box

Or—the grand master's endgame—we can turn those squares into *circles*. Never mind that actually doing this with pencil and paper is called "squaring the circle," and is, in hands-on terms, impossible. "Hands-on" is for gymnasts; we're gliding through the clouds of pure geometry. So just imagine each square slowly morphing into a circle, its area never changing.

Our pyramid becomes a cone. Our cube becomes a cylinder. And thus, a cone fills 1/3 of the cylinder that contains it.

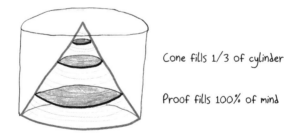

Cone fills 1/3 of cylinder

Proof fills 100% of mind

Pretty cool, right? In the second century, Plutarch gushed:

> *It is not possible to find in all geometry more difficult and intricate questions, or more simple and lucid explanations... No amount of investigation of yours would succeed in attaining the proof, and yet, once seen, you immediately believe you would have discovered it; by so smooth and so rapid a path [Archimedes] leads you to the conclusion required.*

Still, these geometric excursions don't exactly scream "military genius." One has to wonder: Where did his Rome-battering war engines come from?

"These machines he had designed and contrived," Plutarch insists, "not as matters of any importance, but as mere amusements in geometry." Peculiar as that sounds, it's a basic pattern of mathematical history. An aimless flight of fancy leads, somehow and someway, to a technological breakthrough down the road.

Although the Romans didn't much appreciate pure mathematical inquiry, they sure appreciated boat-crushing death claws. Recognizing themselves as the villains in an ancient prequel to *Home Alone*, General Marcellus and his forces retreated.

One afternoon, several months later, Archimedes was scratching diagrams in the dust. I like to imagine he was revisiting his favorite proof—the theorem that he instructed friends and family to commemorate upon his tomb.

It begins with a sphere.

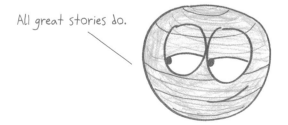

We encase it in a cylinder, so that the fit is perfect and snug, like a single serving of tennis ball.

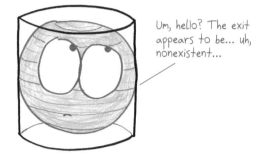

Archimedes's question was this: **What fraction of the cylinder does the sphere fill?**

(Really, his question was more elemental: How big is a sphere? But any description of size requires reference to something already known—e.g., my height is approximately $5\frac{2}{3}$ of those preexisting units called feet—and that's where the cylinder comes in.)

To begin, cut the whole shape in half. Instead of a tennis ball in a snug container, we've got a hemisphere in a hockey puck.

Now, instead of worrying about the volume *inside* the hemisphere, we're going to focus on the volume *outside* of it. In the spirit of Infinite Mischiefs, we can think of this region as a stack of hoops or washers. Each is a circle with a circular hole cut out.

At the bottom of this stack is a superthin washer. Its hole occupies the whole circle, leaving just a stringlike ring. At the top, meanwhile, is a superthick washer. It's pretty much an intact circle with a pinprick aperture. In between is a whole family of intermediate washers.

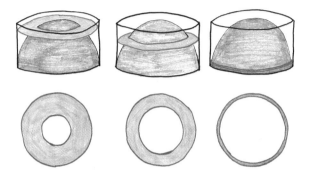

What are the areas of these shapes? With an interlude of slick algebra, we deduce that each has an area of πh^2, where h is its distance above the ground.

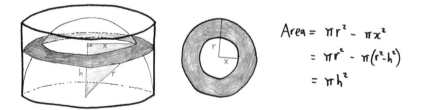

$$\text{Area} = \pi r^2 - \pi x^2$$
$$= \pi r^2 - \pi(r^2 - h^2)$$
$$= \pi h^2$$

This means, invoking the Principle of Infinite Mischiefs, that each can be replaced with a circle of radius h.

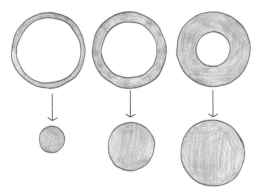

Behold! What remains is no longer a strange hemisphere-shaped crater, but instead, a simple upside-down cone.

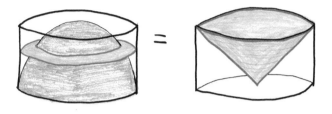

As we've already established, the cone fills 1/3 of the cylinder. Thus, the empty space—i.e., what used to be the hemisphere—fills 2/3.

Conclusion: the sphere fills 2/3 of the cylinder.

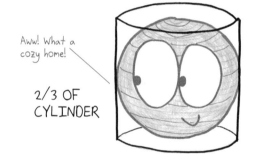

Aww! What a cozy home!

2/3 OF CYLINDER

With these diagrams in the Sicilian sand, Archimedes was dreaming of integrals, millennia in advance. Areas and volumes, infinite slices, rearrangements that solve problems of continuity and curvature: these are the chemical ingredients, the primordial soup, from which the integral would later develop. Why, then, would the world wait so long for calculus's birth?

That day, Rome breached the city. Within hours, Syracuse burned, and soldiers ran amok, looting and killing. "Many brutalities were committed in hot blood and the greed of gain," wrote the historian Livy. Still, the Roman leader Marcellus insisted that the great geometer be spared, "putting almost as much glory in saving Archimedes," says another historian, "as in crushing Syracuse."

Archimedes didn't even notice the city's fall. What's a little plunder and destruction, compared to the engrossing beauty of a figure in the dust?

Historians differ on what Archimedes said when a Roman soldier confronted him. Perhaps he begged: "Please, do not disturb my circles." Perhaps he blustered: "Stand away, fellow, from my diagram." Perhaps he shielded the dust with his hands, as if his ideas were more precious than his life: "Let them come at my head, but not at my line!" In any case, sources agree that the soldier slew him. His blood filled the grooves that his fingers had traced. General Marcellus insisted on a proper burial, and honored Archimedes's relatives with gifts and favors. But the man of Infinite Mischiefs was dead.

Today, Archimedes's greatest legacy lies not in catapults and claws, but in geometry. His lucid arguments, his grasp of infinity, how close he came to calculus. Could a little extra nudge have brought him there? Could calculus have emerged on Earth millennia earlier than it did?

Consider the testimony of mathematician Alfred North Whitehead:

> *The death of Archimedes by the hands of a Roman soldier is symbolical of a world-change of the first magnitude: the Greeks, with their love of abstract science, were superseded in the leadership of the European world by the practical Romans.*

Nothing wrong with practicality. Or is there? Nineteenth-century British prime minister Benjamin Disraeli defined a practical man as one "who practices the error of his forefathers." According to Whitehead, that's just what the Romans did. Nowhere in the vanquishing civilization could you find the imaginative spark of the vanquished.

> *All their advances were confined to minor technical details of engineering. They were not dreamers enough... No Roman lost his life because he was absorbed in the contemplation of a mathematical diagram.*

Centuries later, when the local Syracusans had all but forgotten Archimedes's legacy, the writer Cicero set out in search of his grave. He found it "hidden by bushes of brambles and thorns": "a little column, just visible above the scrub." He recognized it by the carving on top, just as Archimedes had requested: a sphere and a cylinder. The grave has long since vanished, but the proof remains etched in our collective imagination—a medium that can outlast dust, or blood, or all the stonework of Roman hands. ∎

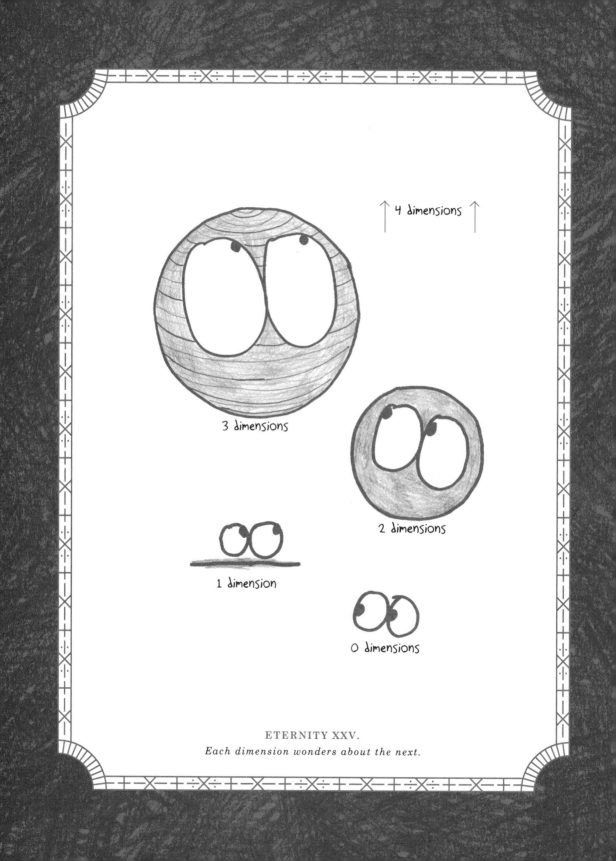

↑ 4 dimensions ↑

3 dimensions

2 dimensions

1 dimension

0 dimensions

ETERNITY XXV.
Each dimension wonders about the next.

XXV.

FROM SPHERES UNSEEN

The world of *Flatland: A Romance of Many Dimensions* lives up to its name. The setting of the classic 1884 novella is flat: flatter than pancakes, flatter than paper, flatter than the female characters in a Michael Bay film. It's a two-dimensional cosmos, endowed with length and width, but no depth. And yet its inhabitants—triangles, squares, pentagons, and the like—sense no missing dimension. In fact, like Kansans in Kansas, or Texans in Texas, they can imagine no existence beyond it.

Until, one day, a very strange visitor arrives.

At first, the sphere looks like it is only a point, appearing out of nowhere. Then, as it passes through Flatland, our narrator (named "A Square") sees a circle, gradually growing in size.

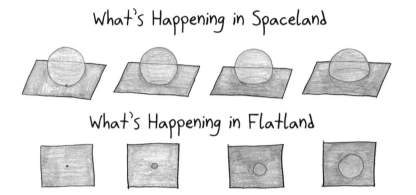

Peculiar indeed! Imagine how you'd feel if some guy strolled through the door and, as he did so, grew in height from 4 feet to 6 feet. (Maybe you'd feel like me every time I teach 9th grade.) A Square inquires what the heck is going on, but receives only cryptic replies, like this one:

> *You call me a Circle; but in reality I am not a Circle, but an infinite number of Circles, of size varying from a Point to a Circle of thirteen inches in diameter, one placed on the top of the other. When I cut through your plane as I am now doing, I make in your plane a section which you, very rightly, call a Circle.*

In this odd circumlocution, the sphere reveals a way to conceive of its nature. A sphere is a stack of infinite disks, varying in radius, each infinitesimally thin. To understand the sphere is to combine—to sum—all these little circles into a single unified whole.

A sphere is an integral of circles.

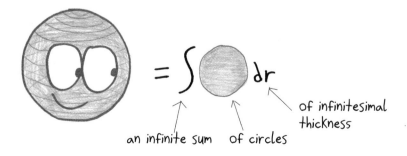

If you've endured a first-year calculus course, then you've encountered this notion before. It's the culminating topic, a piece of mathematics that puts a special spin on the idea of solidity.

(Warning: if you're prone to motion sickness or bad puns, the emphasis here is on "spin.")

To start, you pick a flat two-dimensional region. Next, rotate it around an axis, like a rigid flag spun rapidly around a pole. The space it passes through will form a 3D object, known as a "solid of revolution":

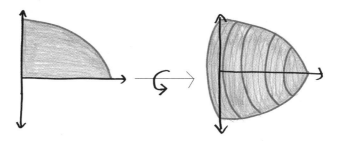

This potter's-wheel process elaborates 2D regions into 3D ones, bringing Flatland into Spaceland. If you want to know the volume of the pottery we've created, the approach is simple. Just analyze the solid as a stack of infinitely thin disks, and then integrate them.

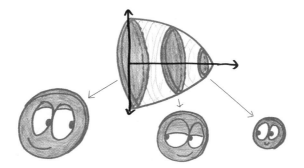

To compute the volume of the spherical intruder, we must first select the appropriate 2D region. What shape, when spun around an axis like a rotisserie chicken, will generate a sphere with a diameter of 13 inches?

Fiddle with your mental 3D printer, and I believe you'll find that a semicircle does the trick.

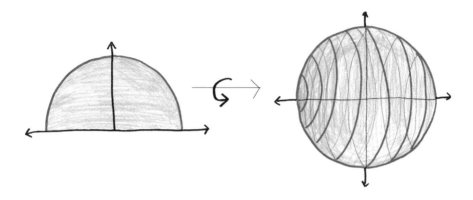

Fun fact about semicircles: they're full of radii. Fun fact about radii: each one forms the hypotenuse of a right triangle. This means that the coordinates of every point on our semicircle obey the Pythagorean theorem.

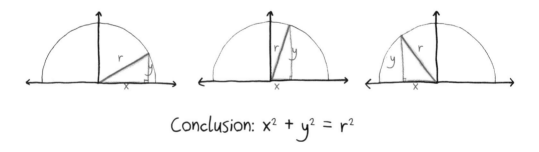

Conclusion: $x^2 + y^2 = r^2$

With a little algebraic manipulation—which, like any considerate host, I have swept under the rug and out of sight—we arrive at the appropriate integral. This will sum together an infinite collection of infinitely thin disks. They'll begin with a radius of zero, grow to a maximum radius of 6.5, and then subside back to a radius of zero. Just like the sphere passing through Flatland.

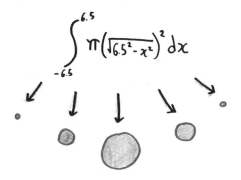

$$\int_{-6.5}^{6.5} \pi \left(\sqrt{6.5^2 - x^2} \right)^2 dx$$

I'll again spare you the algebraic details and deliver you straight to our final result, the volume of the mysterious sphere: $\frac{4}{3}\pi\,6.5^3$ or approximately 1150 cubic units.

In two consecutive chapters, we've calculated the size of a sphere. Perhaps you've noticed commonalities. Both methods begin by cutting the problem in half; both involve an infinite dissection; both yield fancy-looking pictures. And yet they leave pretty different tastes in your mouth, don't they? Myself, I prefer Archimedes's argument. It's slick. It's clever. It fits together just so. A work of craftsmanship, ingenuity, even art.

As for the "solid of revolution" approach—well, I can't say it satisfies the soul. After a promisingly aesthetic beginning (the spinning! the infinite layers!), it ends with several lines of brute algebra. It's like a hike that somehow leads from a scenic hilltop into an airport terminal. An elegant puzzle is thus reduced to a technical exercise.

And that's exactly the point.

We can't all be Archimedes. In fact, statistics indicate that none of us are Archimedes. If we rely on cosmic blasts of cleverness to solve our problems, we'll be waiting for eons. To get anything done, we need to convert the mystical into the mechanical, the fluid into the fixed, the ineffable into the plainly effable.

Solids of revolution embody this spirit. All of us can now safely walk where before only Archimedes could tread. That's the whole point of calculus: to give a systematic approach to otherwise daunting puzzles. To make each of us into an autopilot Archimedes. Vast populations of shapes can be dissected and understood via solids of revolution, from cubes to cones to pyramids to Mickey Mouse dolls.

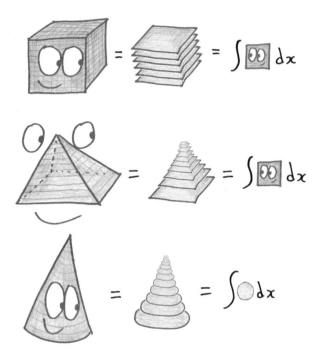

How might all this look to our Flatlander hero, the intrepid and rather overwhelmed A Square? Remember, at the story's beginning, he can't see the third dimension. Can't even imagine it. Consider his account of what Flatland visual life is like:

> *Place a penny on the middle of one of your tables in Space; and leaning over it, look down upon it. It will appear a circle.*

> *But now, drawing back to the edge of the table, gradually lower your eye (thus bringing yourself more and more into the condition of the inhabitants of Flatland), and you will find the penny becoming more and more oval to your view; and at last when you have placed your eye exactly on the edge of the table (so that you are, as it were, actually a Flatlander) the penny will then have ceased to appear oval at all, and will have become, so far as you can see, a straight line.*

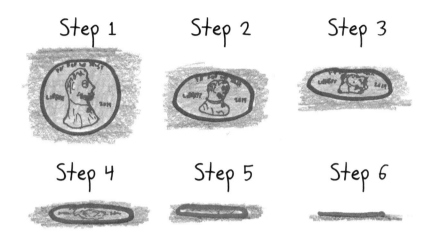

As 3D creatures, we see in 2D; our field of vision is like a painter's canvas or a movie screen. By the same token, Flatland's 2D creatures see in 1D. Their visual field is a naked horizon, with nothing above or below.

How, then, would you explain the third dimension to the poor guy? In the story, the sphere's attempts go nowhere:

> *I. Would your Lordship indicate or explain to me in what direction is the Third Dimension, unknown to me?*

> *STRANGER. I came from it. It is up above and down below.*

> *I. My Lord means seemingly that it is Northward and Southward.*

> *STRANGER. I mean nothing of the kind. I mean a direction in which you cannot look... in order to see into Space you ought to have an eye, not on your Perimeter, but on your side, that is, on what you would probably call your inside; but we in Spaceland should call it your side.*

> *I. An eye in my inside! An eye in my stomach! Your Lordship jests.*

When language and intuition fail, we have just one recourse remaining. No, not "mouthfuls of hallucinogens." I mean calculus. Even if A Square cannot conceptualize the visitor's shape, he can nevertheless calculate his volume. Computing an integral requires no in-depth visualization or first-hand experience. Just technical proficiency.

When in doubt, math it out.

In college, a friend introduced *Flatland* to me as "the closest you'll ever come to actually seeing the fourth dimension." The pivot comes toward the end of the book, when A Square asks the sphere to show him not just 3D, but 4D:

> *As you yourself, superior to all Flatland forms, combine many Circles in One, so doubtless there is One above you who combines many Spheres in One Supreme Existence, surpassing even the Solids of Spaceland. And even as we, who are now in Space, look down on Flatland and see the insides of all things, so of a certainty there is yet above us some higher, purer region...*

The Sphere refuses his own medicine. "Pooh!" he cries. "Stuff! Enough of this trifling!"

I have to confess my sympathy with the sphere. If there were a fourth spatial dimension, then our 3D reality would constitute an infinitely thin slice of it. A visitor from 4D would appear out of nowhere, in the middle of the room, and proceed to fluctuate in size, because we'd see only a single cross section at a time—not the creature itself, but an infinitesimal layer thereof.

I can say all that. But, sure as syrup on pancakes, I cannot picture it.

Still, where cleverness cannot take root, calculus sometimes can. To compute the volume of a 4D sphere, I just need to integrate an infinite collection of 3D ones:

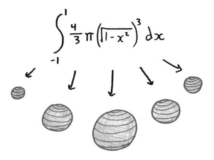

$$\int_{-1}^{1} \frac{4}{3}\pi \left(\sqrt{1-x^2}\right)^3 dx$$

This one took me a full page of algebra, plus several tweets' worth of coaching from kind internet pals. But I got there.

I'm reminded of mathematician Steven Strogatz's account of his own school days:

> *I was a grinder.... My style was brutal. I'd look for a method to crack the problem. If it was ugly or laborious, with hours of algebra ahead, I didn't mind because the right answer was guaranteed to emerge at the end of the honest toil. In fact, I loved that aspect of math. It had justice built into it. If you started right and worked hard and did everything correctly, it might be a slog but you were assured by logic to win in the end. The solution would be your reward.*

> *It gave me great pleasure to see the algebraic smoke clear.*

That's the gift calculus gives us. It validates our sense of cosmic justice, our faith in the grind, our confidence that those long hours of toil will, in the end, bring victory. In this case, when the smoke clears, and you finally gaze upon the volume of the 4D hypersphere, it turns out to be the rather adorable $\frac{\pi^2}{2}$.

The units, please note, are quartic meters: a meter to the fourth power. What that means, I don't know, but I bet Archimedes didn't either. And somehow, that consoles me. ■

ETERNITY XXVI.
David Foster Wallace makes an infinite gesture.

XXVI.

A TOWERING BAKLAVA
OF ABSTRACTIONS

This is a chapter about a two-page endnote published in 1996. Perhaps that sounds arcane, so let me dispel any doubt: it *is* arcane. Fantastically so. The endnote in question imports a prickly, cactus-like topic from one arid setting to another—from the desert of introductory calculus to the bizarre greenhouse of experimental fiction. The book in which the endnote appears—*Infinite Jest* by David Foster Wallace—has been dubbed "a masterpiece," "forbidding and esoteric," "the central American novel of the past thirty years," and "a vast, encyclopedic compendium of whatever seems to have crossed Wallace's mind."

My question is this: Why, in a work of fiction, a dream of passion, would Wallace force his soul to this odd conceit? Why devote two breathless pages to—of all things—the mean value theorem for integrals?

What's the MVT to him, or he to the MVT?

Despite the regal name, the mean value theorem is a pretty simple statement. Imagine you've got a quantity changing over a time period—rising, falling, falling, rising. The MVT asserts that somewhere amid the flux and flow, there's a magical moment: an instant when the value is equal to its overall average (or "mean").

Take, for example, a road trip. You travel 200 miles in four hours, your speed fluctuating all the while. The average speed, if you run the calculation, works out to 50 miles per hour.

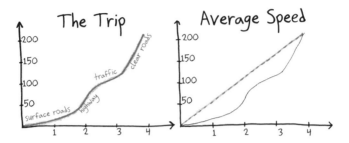

The MVT assures us that—for at least one shining instant during your journey—you were traveling *exactly* 50 miles per hour.

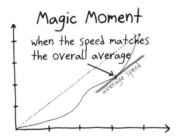

It's simple logic, really. Did you stay above 50 miles per hour the whole four hours? No—that'd bring you more than 200 miles. Did you stay below 50 miles per hour? Again, no—that'd bring you less than 200. Did you leap straight from below 50 miles per hour to above 50 miles per hour, never passing through 50 miles per hour itself? Not unless you drive a souped-up DeLorean. We thus conclude that—for at least one moment—you were moving precisely 50 miles per hour.

Another example: Say the temperature is fluctuating throughout the day. It goes up. It goes down. It goes back up. You make awkward small talk about it, because "discuss the weather" is the default home page in your social programming.

Now, how do we find the "mean" (or average) temperature?

Well, to average a few numbers, we sum them up, then divide by the size of the data set. If you've scored 70, 81, and 89 on the last three tests, then your average is the total (240) divided by the size of the data set (3). That gives you 80. But with temperature, there are *infinitely many* numbers, one for each moment of the day. To sum them all, we need an integral.

COMPUTING AVERAGES

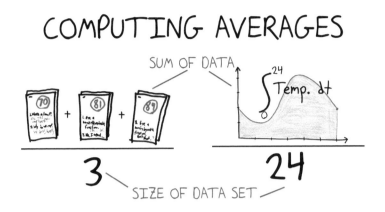

Notice that, in the diagram below, the integral is smaller than the rectangle on the left and larger than the rectangle on the right, just as the average temperature is smaller than the maximum but larger than the minimum.

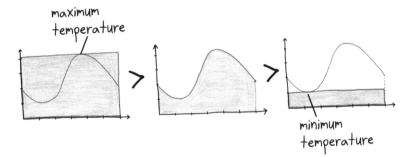

What does the MVT tell us? Simply that, at some moment during the day, the temperature equals the average.

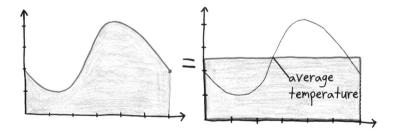

So much for the MVT. Now we turn our attention to DFW, David Foster Wallace himself, to see what he makes of this technical little theorem. On page 322 of *Infinite Jest*, we encounter a "complicated children's game" called Eschaton. It requires "400 tennis balls so dead and bald they can't even be used for service drills anymore," each representing a thermo-nuclear warhead. Players divide into teams (representing global actors), then receive their allotment of warheads, computed via the MVT for integrals.

An endnote—I should say, *the* endnote—directs us to page 1023. Here, we learn that for each nation, the relevant statistic for sizing the nuclear arsenal is $\frac{\text{GNP} \times \text{Nuclear Spending}}{(\text{Military Spending})^2}$. The higher this figure, the greater the nuclear firepower. But, rather than apportion tennis-ball nukes on the *current* value, Eschaton operates on a *moving average* of this value over the last several years, a computation which (according to Wallace's narrator) requires the MVT.

Now, if none of this is making any sense to you, fear not. The fact is that none of this makes any sense to anyone.

The MVT is what's called an "existence theorem." It tells us that the temperature must, at some moment, achieve its daily average. It does not—it cannot—tell us where or when this moment occurs. It only gestures toward the haystack of time, and assures us of the needle's presence, somewhere among those infinite moments.

I had to read Wallace's endnote several times to convince myself that, yes, he's invoking the MVT for the purposes of computation, and, no, that doesn't work. I haven't even discussed the ill-chosen statistic (why punish countries for nonnuclear military spending?) or the fake-news explanation of the MVT (insisting that only the minimum and maximum are needed to compute the mean). The whole passage is Jude Wanniski–level claptrap.

Which only intensifies my original question: Why, DFW, why?

According to Wallace's own writing, mathematics is the thread that holds his life narrative together. "As a child I used to cook up what amount to simplistic versions of Zeno's Dichotomy," he once wrote, "and ruminate on them until I literally made myself sick." Even his gift for tennis boiled down to mathematics. "I was regarded as a kind of physical savant," he wrote, "a medicine boy of wind and heat...sending back moon balls baroque with ornate spins." Wallace remembered his midwestern home (Urbana, Illinois) as a giant Cartesian plane:

> I'd grown up inside vectors, lines and lines athwart lines, grids—and, on the scale of horizons, broad curving lines of geographic force.... I could plot by eye the area behind and below these broad curves at the seam of land and sky way before I came to know anything as formal as integrals or rates of change. Calculus was, quite literally, child's play.

But as an undergrad at Amherst College, he stumbled over his first mathematical hurdle. "I once almost flunked a basic calc course," he wrote, "and have seethed with dislike for conventional higher-math education ever since." He elaborated:

> The trouble with college math classes—which...consist almost entirely in the rhythmic ingestion and regurgitation of abstract information...is that their sheer surface-level difficulty can fool us

into thinking we really know something when all we really "know" is abstract formulas and rules for their deployment. Rarely do math classes ever tell us whether a certain formula is truly significant, or why, or where it came from, or what was at stake.

I've met students who share that frustration. It drives most of them to seek concrete examples. DFW being DFW, he sprinted in the opposite direction, toward the discipline's most stupefying and abstract corners. "It is in areas like math and metaphysics," Wallace gushed, "that we encounter one of the average human mind's weirdest attributes. This is the ability to conceive of things that we cannot, strictly speaking, conceive of." As mathematician Jordan Ellenberg observes, "He was in love with the technical and analytic."

After becoming a professional writer, Wallace kept circling back to mathematics. In one interview, he explained that *Infinite Jest* borrows its structure from a notorious fractal called the Sierpinski gasket.

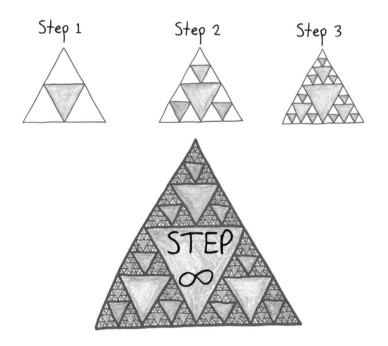

The "DFW + math" love affair culminated in his book *Everything and More: A Compact History of Infinity*. It's a dense, technical treatise on Wallace's favorite branch of modern mathematics: Cantor's theory of infinity.

If the title *Infinite Jest* didn't make it clear, Wallace adored infinity:

> *It's sort of the ultimate in drawing away from actual experience. Take the single most ubiquitous and oppressive feature of the concrete world—namely that everything ends, is limited, passes away—and then conceive, abstractly, of something without this feature.*

I read *Everything and More* back-to-back with Eugenia Cheng's *Beyond Infinity*. It made for a funny juxtaposition. Cheng—a research mathematician—chose to write a breezy, nontechnical piece of pop science, brimming with friendly analogies. Wallace—a novelist—chose to cultivate a thornbush of forbidding notation. It felt rather backward. "One wonders exactly whom Wallace thinks he is writing for," mused philosopher David Papineau in a *New York Times* review. "If he had cut out some of the details, and told us rather less than he knows, he could have reached a lot more readers."

That's the knock on DFW. He never, *ever* told you less than he knew.

It seems to me that DFW was attracted to math, in large part, by the same qualities that repel others—perhaps even *because* they repel others. "Modern math is like a pyramid," he wrote, "and the broad fundament is often not fun…math is perhaps the ultimate in acquired tastes."

Consider, for example, an elder cousin of the MVT: the intermediate value theorem. My students tend to consider it a piece of blinding obviousness dressed up in mathematical verbiage. In human terms, it says that if you're five feet tall one year, and five three a year later, then sometime in between, you must have been five one.

No surprises there.

As presented in textbooks, it says that if a function f is continuous for all values x where $a \le x \le b$ and either $f(a) \le k \le f(b)$ or $f(b) \le k \le f(a)$, then there exists some c such that $a \le c \le b$ and $f(c) = k$.

Why this tsunami of symbols to express such a no-duh notion?

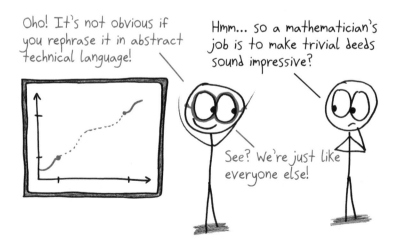

Well, in the 19th century—the period DFW explores in *Everything and More*—new questions of infinitude began to rattle mathematicians. Which sums converge to reasonable answers? Which sums don't? What do we really know, and how do we know it? With painstaking caution, a community of mathematicians sought to rebuild calculus from the ground up: anchoring it not in geometry or intuition, but in arithmetical inequalities and precise algebraic statements. This is when the IVT—as well as the MVT—became a thing. If you want to prove every fact in calculus, step by step, then these theorems are indispensable.

But is this the only true "mathematics"? Were all those earlier generations, from Archimedes to Liu Hui to Agnesi, just stumbling toward our "correct" notions of analytic rigor, like ancient heathens in Dante's purgatory, biding their time until the birth of Christ?

When DFW celebrates math, he means a certain kind of math: a kind born in the 1800s for contingent historical reasons; a kind closer to analytic philosophy (Wallace's undergraduate major) than to rich veins of geometry, combinatorics, and more; a kind Wallace dubs "a towering baklava of abstractions," capturing the sweet nuttiness of the endeavor but not its baffling meaninglessness to the many students who grow shy and queasy at such dense technicality; a kind that Wallace parades, error-ridden, through a novel where its only conceivable role is to obfuscate and impress; a kind that I enjoyed in college but have since drifted away from, feeling none the poorer for it, because its peculiar aesthetic is not the only one a mathematician can admire.

Math is a weave of many threads: the formal and the intuitive, the simple and the profound, the momentary and the eternal. Love the thread you love. But never mistake it for the tapestry. ∎

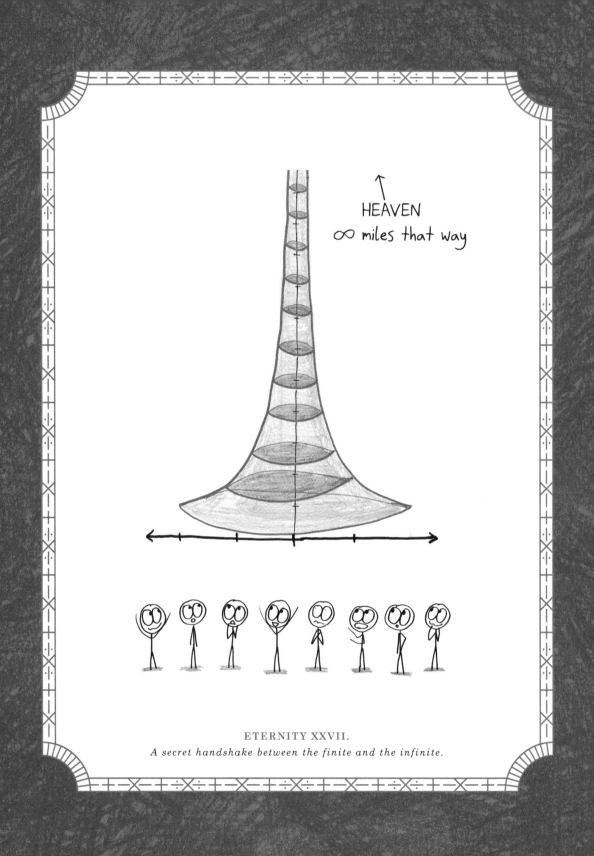

HEAVEN
∞ miles that way

ETERNITY XXVII.
A secret handshake between the finite and the infinite.

XXVII.
GABRIEL, BLOW YOUR TRUMPET

An old riddle asks whether God, being omnipotent, can create a stone so heavy that even God cannot lift it. The question is a theological trap. Say no, and you undermine God's powers of creation; say yes, and you dis God's upper-body strength. The word for this is "paradox," a wound that logic inflicts upon itself. It's an argument whereby seemingly correct assumptions lead by seemingly correct logic to seemingly bananas conclusions.

And if you think theology breeds paradoxes, just wait until you meet mathematics.

"Gabriel's horn," one of my favorite paradoxes in calculus, draws its name from the archangel Gabriel. His trumpet, which rattles Earth with messages from heaven, is wondrous and frightening, finite and infinite, a bridge between the mortal and the divine. It's a fitting name for an object of inherent contradiction.

To craft the horn, first draw the curve whose equation is $y = \frac{1}{x}$. As the distance x grows, the height y descends. When x is 2, y is $\frac{1}{2}$. By the time x is 5, y has fallen to $\frac{1}{5}$. And so it goes, all down the axis.

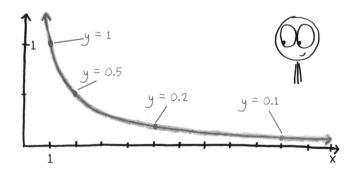

Before long, x grows quite large and y quite tiny. When x is a million—which should occur about 6 miles down the road—then y falls to $\frac{1}{1,000,000}$, about the thickness of a cell membrane.

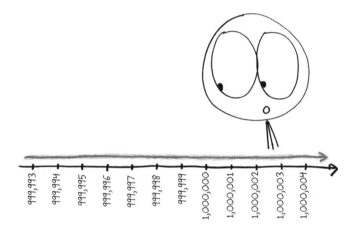

By the time x is a billion—which, if you're holding this book in Los Angeles, should occur near Moscow—y is $\frac{1}{1,000,000,000}$. By my calculations, that's about half the width of a helium atom.

Still the curve runs on, never reaching the axis, toward the numberless horizon we dub "infinity."

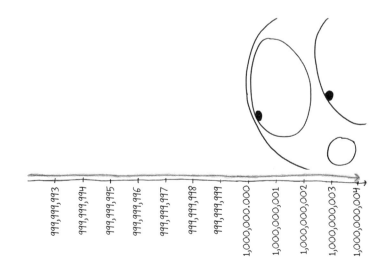

Next, we need to rotate this curve around the *x*-axis, solid-of-revolution style, to yield a 3D figure. This spindly beauty—a collection of infinitely many disks, each infinitesimally thin—is Gabriel's horn.

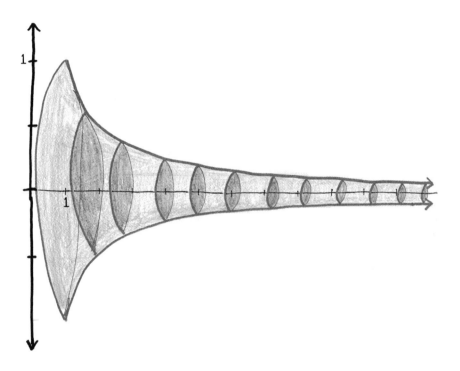

Like any 3D object, it permits two kinds of measurement. First, we can measure its *volume*—i.e., how many cubic units of water you'd need to fill it. And second, we can measure its *surface area*—i.e., how many square units of wrapping paper you'd need to cover it.

Volume first. In physical reality, an infinite object can't have finite volume; you'd need a horn that grows thinner than an atom, which not even the deftest fine motor skills can manage. But math inhabits a different kind of reality, where such feats are routine. So deploying standard methods, we set up the integral $\pi \int_{1}^{\infty} \frac{1}{x^2}\,dx$, which resolves to π. Thus, the volume of Gabriel's horn is 3.14 cubic units, give or take.

Next up: surface area. The integral comes out a little uglier: $2\pi\int_1^\infty \frac{1}{x}\sqrt{1+\frac{1}{x^4}}dx$. But that's just slightly larger than a far less gnarly integral: $2\pi\int_1^\infty \frac{1}{x}dx$. This turns out to equal…well, no particular number. Its finite approximations grow without bound. And since the surface area is a little bigger than this, we conclude that Gabriel's horn has a surface area of ∞.

We're now on the doorstep of contradiction. Gabriel's horn has finite volume—so if you wanted to fill it with paint, you could. And yet, Gabriel's horn has infinite surface area—so, if you wanted to paint the surface, you couldn't.

But…once you fill it with paint, isn't every point on the surface now painted?

How can these both be true at once?

The first to explore this paradox figure was 17th-century Italian Evangelista Torricelli. Along with pals Galileo and Cavalieri, he was blazing "the Royal Road through the mathematical thicket," using the newfangled mathematics of infinitesimals. "It is manifest," wrote Cavalieri, "that plane figures should be conceived by us like cloths woven of parallel threads; and solids like books, composed of parallel pages."

These scholars trafficked in infinite sums, infinitely thin elements, and bizarre objects like Gabriel's horn (which is also known as "Torricelli's trumpet").

This was calculus, kicking in its womb.

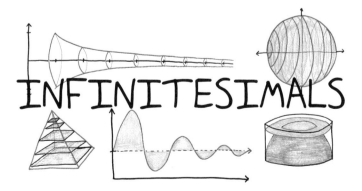

At the time, the Jesuit order was building an admired system of universities across Europe. More than just good schools, these were *Catholic* schools. "For us," said one leader, "lessons and scholarly exercises are a sort of hook with which we fish for souls." In this curriculum, math played a central role. "Without a doubt," said a Jesuit named Clavius, "the mathematical disciplines have the first place among all others."

But not just any math: it had to be Euclid. Euclidean geometry advanced via clear logic, from self-evident assumptions to ironclad conclusions, with nary a snag or a paradox. "The theorems of Euclid," said Clavius, "retain... their true purity, their real certitude." The Jesuits saw Euclid as a model for society itself, with the pope's authority as the irrefutable axiom.

As for Torricelli's work, the Jesuits were not fans. Historian Amir Alexander, in his book *Infinitesimal: How a Dangerous Mathematical Theory Shaped the World*, explains: "Whereas Euclidean geometry was rigorous, pure, and unassailably true, the new methods were riddled with paradoxes and contradictions, and as likely to lead one to error as to truth." The Jesuits saw Gabriel's horn as anarchist propaganda, an anathema to order. "Theirs was a totalitarian dream of seamless unity and purpose that left no room for doubt or debate," writes Alexander. As Ignatius, another Jesuit of the time, put it: "What I see as white I will believe to be black if the hierarchical Church thus determines it."

So the pope banned the infinitesimal. Torricelli became a mathematical outlaw. Gabriel's horn was intellectual contraband.

The irony is that this paradox is not so hard to resolve. How can Gabriel's horn have an interior that paint can fill but an exterior that paint cannot cover? It all depends on how we think about paint.

As mathematician Robert Gethner explains, the paradox rests on an assumption: that "surface area" corresponds to "paint required." But paint isn't two-dimensional. "If we are planning to paint a room...," he writes, "we wouldn't ask for 1000 square feet of paint." Like paper, paint is three-dimensional. It has a thickness, albeit a small one.

So, one approach: allow the paint's thickness to diminish, growing thinner and thinner as Gabriel's horn narrows. Under this assumption, it's possible to cover the surface with a finite quantity of paint. Paradox resolved.

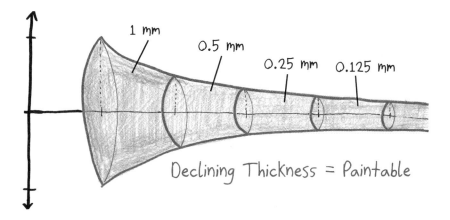

Or, if you prefer, take another approach: assume the paint needs some minimum thickness. (That's more how physical paint is; for example, you can't slather a layer $\frac{1}{1000}$ of an atom thick.) Thus, way down the axis, the horn thins to subatomic scales, but the paint does not. Eventually, the paint looms trillions of times thicker than the thing it's painting. This restores our original finding that it's impossible to paint the horn—except now, it's also impossible to *fill* the horn, because at some point, it becomes thinner than the thinnest possible paint.

Under this assumption, the horn can be neither painted nor filled. Again, paradox resolved.

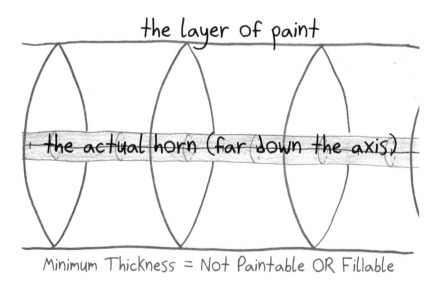

the layer of paint

the actual horn (far down the axis)

Minimum Thickness = Not Paintable OR Fillable

Whether the Jesuits of the 1600s made a religious error, I have no standing to say. But I believe they made a mathematical one. Paradox is not something to be feared or eradicated. It's a spur to thought, an invitation to contemplation.

Paradoxes arise not just in the musty halls of theology and mathematics, but also, according to business professor Marianne Lewis, in corporate settings. Elements that "seem logical when considered in isolation"—a short-term target, a long-term vision, a strategic priority—become

"irrational, inconsistent, even absurd, when juxtaposed." That's not necessarily a bad thing. "Paradoxes provide creative friction," she writes.
"Understanding paradox may hold a key to coping with, and even excel-
ling in the face of, strategic tensions." Paradox is the grain of sand that
helps form the pearl of theory.

Douglas Hofstadter, author of *Gödel, Escher, Bach*, goes further. "The
drive to eliminate paradoxes at any cost..." he writes, "puts too much stress
on bland consistency, and too little on the quirky and bizarre." Paradoxes
are pleasures in their own right, M. C. Escher paintings without the paint.
Or, in this case, with infinite paint. ∎

ETERNITY XXVIII.
A fool ages; a beautiful integral does not.

XXVIII.

SCENES FROM AN IMPOSSIBILITY

My first brush with the Impossible Integral came in the spring of 10th grade.

The 12th graders were crowded into the lobby. Their arms were an octopus of pens, covering a poster with doodles, signatures, and out-of-context quotes from physics teacher Mr. Riahi. It was a sprawling collage, full of jokes I didn't get, jokes I half got, jokes I wanted so badly to get.

Amid the chaos, I spotted a peculiar jumble of symbols:

I pointed. "What's that?"

"It's the integral of e to the negative x squared," explained David, explaining nothing.

"So," I asked, "what's the joke?"

"THE JOKE IS, IT'S IMPOSSIBLE," replied Abby, who tended to speak in block capitals.

"Like…was it on a test? Nobody could solve it?"

They chuckled.

"AH, SOPHOMORE BEN THREE OF THREE," Abby addressed me. (It was true: Ben Copans and Ben Miller came before me, alphabetically.) "HOW INNOCENT. HOW TOUCHING."

"Though, if by 'on a test' you mean 'in the universe,'" mused Bart, "then yes. It was on a test. And nobody could solve it."

"So," I said, "it's like…dividing by zero?"

"More like squaring the circle," answered David.

"IT'S LIKE RAIN ON YOUR WEDDING DAY," Abby elaborated. "IT'S LIKE 10,000 SPOONS WHEN ALL YOU NEED IS A KNIFE."

Abby wasn't wrong. Way back in the infancy of calculus, Johann Bernoulli remarked on the specter of impossible integrals. "Sometimes," he wrote, "we cannot say with certainty whether the integral of a given quantity can be found or not." In the 19th century, analyst Joseph Liouville said it with certainty: some integrals cannot be found. Take $\int \sqrt[4]{1 + x^2}\, dx$, or $\int \ln(\ln x)\, dx$, or $\int \frac{1}{\arctan(x)}\, dx$. These lack clean solutions. The phrase is "unsolvable in elementary terms." Summon all the sines, all the cosines, all the logarithms and cube roots you like; no assemblage of the standard algebraic parts will ever yield a formula. It's a lock with no key, a riddle with no solution, a tough piece of steak in a world of 10,000 spoons.

I gazed at the symbols. The big squiggle didn't mean anything to me, not yet. "We start calculus in nine months," my friend Roz had remarked earlier that term, "and you know what that means: someone is pregnant with my graphing calculator." Roz's joke I got, but the seniors' one eluded me.

Fast-forward eight years.

My first teaching gig led me to a former car dealership on the edge of Oakland's Chinatown neighborhood. It was one day during my third year that I showed my AP Calculus students the Impossible Integral: $\int e^{-x^2}\, dx$. With a rhetorical flourish, I revealed its impossibility.

"Spoilers!" Adriana cried.

A few students shook their heads.

But Betsaida pressed me: "So...there's no area under the curve?"

Ah—a good question. As will surprise no one, I'd been sloppy and unclear. It turns out that the function e^{-x^2} has a perfectly nice graph:

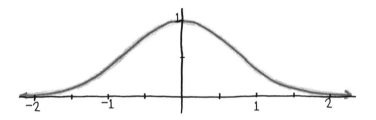

If you seek a specific enclosed area—0 to 1, say, or 0.9 to 1.3, or −1.5 to 0.5—then there is indeed an answer.

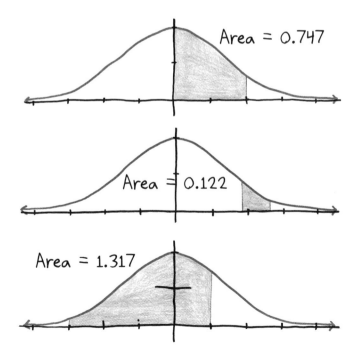

So why "impossible"? Because there's no good formula for those areas. Our magic wand—the fundamental theorem of calculus, which computes integrals by taking antiderivatives—proves here to be a powerless twig. Wave the wand for eons; no magic answer will ever issue.

"My graphing calculator can do it," said Yu Hang. "So, according to you, it's smarter than all the mathematicians in the world."

"Well," I scoffed, "it's definitely faster at *approximating* integrals via Riemann sums." That's the best we can do with such functions: approximate. My tone conveyed my prejudice. An approximation isn't a *real* solution. It's rough. Second-rate. Doesn't count.

"Are you sure?" Yu Hang goaded. "The calculator seems smarter to me."

When the lesson ended, I slipped through the door in the back, which led Narnia-style to the Statistics classroom. It was my first year teaching Stats, and I was stumbling. My instincts were all pure math, all proof and abstraction—and all wrong for statistics. I felt as if I were mixing sports, like teaching the kids to bowl by thwacking tennis balls at the pins.

"For adult men in the US," I began, "the average height is 70 inches, and the standard deviation is 3.5 inches. How many men do we expect to be at least 7 feet tall?"

Like so much in statistics, this problem rests upon the gentle slope of the normal distribution, a.k.a. the bell curve. Its ubiquitous shape describes measurement errors, diffused particles, IQ scores, amounts of rainfall, large numbers of coin flips—and, to a decent approximation, human heights. So in we dove.

Step 1: 7 feet is 84 inches.

Step 2: That's 14 inches above the average.

Step 3: That's 4 standard deviations above the average.

Step 4: We flipped to the table in the back of the textbook, and learned that 4 standard deviations corresponds to a percentile of... Um, actually, the chart stops at 3.5 standard deviations. Oops.

Step 5: Apologizing, I pulled out my laptop and opened Excel, which could carry us beyond the table's edge. Our answer: 0.999968. In other words, a 7-foot man in the US sits at the 99.9968th percentile—roughly 1 in 30,000.

We were dissecting this result—which is to say, debating the relative merits of Shaquille O'Neal vs. Yao Ming—when it occurred to me that, without plan or intention, I'd taught two consecutive and wildly divergent lessons on the exact same topic.

You see, the normal distribution we're using looks like this:

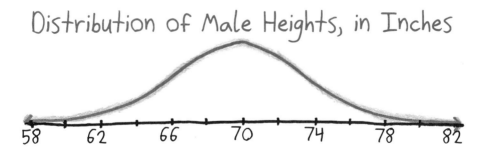

It's just a massaged version of e^{-x^2}: shifted and flattened, but the same function at heart. Which means that it has no integral. And yet, there we were, integrating it. Daily. Constantly.

Heedless of impossibility, the whole discipline of statistics rests upon integrating what cannot be integrated. Every segment of the population (those between 5′11″ and 6′2″, for example) corresponds to an area under this curve:

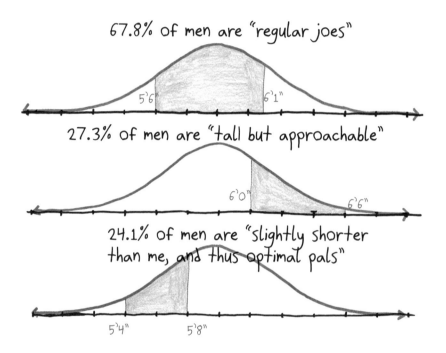

Nature doesn't worry about symbolic antiderivatives. The tables in the textbook, the built-in formula in Excel, the graphing calculator in Yu Hang's backpack—these tools give solid numerical approximations. Most days, that's all you need. As Albert Einstein put it: "God does not care about our mathematical difficulties. He integrates empirically."

I stood there, whiteboard marker in hand. My lightning flash of insight was followed, as always, by the slow, rumbling thunder of regret. I wanted to throw open the door that divided Calculus from Statistics. I wanted to explain what a fool I'd been, to confess my pure-math chauvinism. "Abstract formulas," I wanted to shout, "are not inherently better than concrete approximations! God integrates empirically, and the truth will set us free!"

What held me back—other than impugning my sanity, or (what amounts to the same thing) interrupting Ms. Fleming's chemistry class—was the thought of Yu Hang smirking. *I told you so,* he'd taunt. *Calculators are smarter than mathematicians.* I couldn't speak for the whole profession, but in my case, I knew he'd be right.

Fast-forward again, to the present day.

I'm at my favorite café, sipping the dark roast and "researching" (i.e., not writing) the book in your hands, when I come across an integral named, like many things that Carl Gauss did not discover, after Carl Gauss:

$$\int_{-\infty}^{\infty} e^{-x^2} dx = \sqrt{\pi}$$

It seems that the Impossible Integral allows an exception. If your area sweeps the entire length of the number line, from the distant left to the distant right, from one unfathomable horizon to its unreachable mirror image, then it *is* possible to find an answer.

It is—of all things—the square root of π.

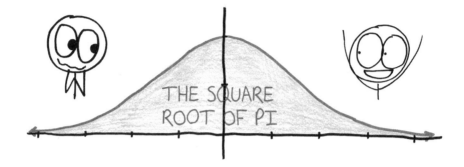

The confluence of fancy symbols (e, π, ∞) reminds me a bit of Euler's identity, $e^{\pi i} + 1 = 0$. That equation gets smothered with adoration from math folks like me. Constance Reid called it "the most famous formula in mathematics"; Ted Chiang describes "a feeling of awe, as if you've come into contact with absolute truth"; Keith Devlin even compares it to "a Shakespearean sonnet that somehow captures the very essence of love."

I can't help but wonder: Where's the rapturous praise for the Gaussian integral? I tweet:

> *I love Euler's identity the way I love the Beatles: with the sheepish knowledge that it probably gets too much attention.*

> *But the Gaussian integral! Look at this beauty, folks! It's the Moody Blues of equations involving* e *and pi!*

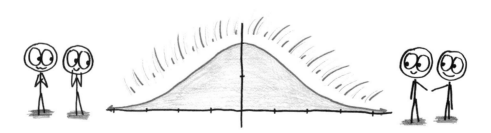

There's just one problem. I don't actually know why this formula is true.

I like to think that I'm a curious learner. I also happen to live with a professional mathematician, whose mind has shelves' worth of knowledge that I lack. But—and here is the paradox of my domestic life—the dots don't connect. I almost never ask her to teach me anything.

Maybe I never got in the habit. Maybe the asymmetries of learning don't jibe with the dynamics of marriage. Maybe I have less curiosity than I prefer to believe, or more stubborn pride. Or maybe, somewhere in the time since 2003, I've ceased being the kid who always scrambles to get the inside joke and become an adult who pretends he already does.

Anyway, today I ask.

Taryn smiles, grabs a coupon for cream cheese, and on the back, shows me how the Gaussian integral is done. Square the whole thing; apply Fubini's theorem; shift to polar coordinates; and, voilà, the result tumbles out. The square root of π.

"So the integral is impossible," I say, "except for the one time that it isn't." Somewhere among those 10,000 spoons, it turns out there's a knife after all.

"Oh," Taryn says, flipping the coupon over. "Do we need cream cheese?"

I get up to check the fridge, and the moment trickles away into eternity. ∎

NARRATOR: They had too much cream cheese.

CLASSROOM NOTES

When I first conceived of this book, I envisioned a linear succession of ideas—pretty much AP Calculus with cartoons. It was the path I had walked as a student and a teacher, the only sensible path, so far as I knew. But the more I followed it, the queasier I became. I wanted a yellow brick road, bursting with color and magic; this felt more like the trail through an Ikea. So, finally taking a hint, I ditched the plan and began to gather stories. Some wove together introductory topics (e.g., the Riemann sum) with advanced ones (e.g., Lebesgue integration). Some themes (e.g., the birth of analysis) became recurring characters, popping up again and again for paragraph-long cameos. Some central topics (e.g., Taylor series) vanished altogether. What I had imagined as a classy string of pearls became a shifting kaleidoscope.

Obviously, this is not a course book. But if you're teaching (or taking) calculus, I hope it can be a friendly travel companion. To help, here is the "standard" sequence of topics (more or less), along with the corresponding stories, and a few stray thoughts on the pedagogical landscape.

Limits: *"What the Wind Leaves Behind" (Ch. 8); "In Literary Circles" (Ch. 16)*

I am by nature a timid, centrist soul: I listen to Coldplay, drink lattes, and have always begun calculus with a unit on limits. But researching this book has radicalized me: I've now fallen in with those hooligans and scofflaws who denigrate limits. Not the mathematical concept—rather, the idea that before meeting derivatives and integrals, the student must undergo a thorough decontextualized study of the local behavior of abstract functions. Next time I teach calculus, I intend to join the rebels by diving right into differentiation, circling back to notions of convergence and continuity only when they arise naturally in context. This is, as I understand it, the path history took, and what's good enough for the Bernoullis is good enough for me.

Tangent Lines: *"Sherlock Holmes and the Bicycle of Misdirection" (Ch. 6)*

Although I've soured on limits as a pedagogical framework, I remain a big fan of the attending philosophical riddles. That's why I love the tangent line problem featured here. It gives a concrete meaning to the notion of "instantaneous motion" and makes for a compelling calculus-free yet calculus-related activity.

Definition of a Derivative: *"The Fugitive Substance of Time" (Ch. 1)*

There is a healthy ongoing debate around how best to introduce the derivative. Is it the slope of a tangent line? The optimal local linearization? The instantaneous rate of change? I've chosen to emphasize the latter perspective, although the "local linearization" idea follows soon, in "When the Mississippi River Ran a Million Miles Long" (Ch. 5).

Differentiation Rules: *"The Green-Haired Girl and the Superdimensional Whorl" (Ch. 10);* "Calculemus!" *(Ch. 15)*

I introduce the derivatives of x^2 and x^3 (in Ch. 10) and the product rule (in Ch. 15) via infinitesimal reasoning. Alas, my newfound radicalism is again rearing its head: where I would have once viewed this approach as illegal, even immoral, I now believe that the philosophical fudgery of imagining dx as "an infinitely small increment in x" is a small price to pay for the enormous pedagogical benefit of bringing geometry into the mix.

Case in point: when I first learned calculus, I was stunned to discover that the sphere's volume ($\frac{4}{3}\pi r^3$), when differentiated with respect to the radius, yielded the sphere's surface area ($4\pi r^2$). Whence this bizarre coincidence? Eventually, I saw the logic: an extra increment of radius grows the volume by a tiny outer layer, effectively equal to the surface area. That fact, which I once viewed as purely algebraic, now feels deeply geometric to me, thanks to the pivotal move of accepting dx and its ilk as (provisionally) meaningful quantities.

Kinematics: *"The Ever-Falling Moon" (Ch. 2);* *"The Fleeting Joys of Buttered Toast" (Ch. 3);* *"Do the Dusty Dance" (Ch. 9)*

In some sense, the canonical derivative is $\frac{dx}{dt}$ = velocity. It is the metaphor by which I understand most (all?) other derivatives. Whenever my students cannot open a particular jar, I rephrase the problem in terms of velocities, and the lid usually pops right off.

The chapter on Brownian motion (Ch. 9) is a case in point. The concept of "differentiability" is elusive in the abstract: Why do we care what can or can't be differentiated? But in the kinematic context, "nondifferentiable" simply means "lacking a speed," which communicates the utter strangeness of the behavior.

By the way, in Ch. 3, James speculates that if his friends knew all of the derivatives of his happiness at this one moment, they could extrapolate the entire lifetime trajectory of his happiness. This presages the Taylor series, in which the derivatives at a single point encode the entire life history of a function.

Linear Approximation: *"When the Mississippi River Ran a Million Miles Long" (Ch. 5)*

Several writers, including Jordan Ellenberg, have persuaded me that "linearization" is the watchword of calculus. I should also confess that I first encountered the Twain passage in Ellenberg's excellent book *How Not to Be Wrong*.

Optimization: *"The Universal Language" (Ch. 4);*
"Princess on the Edge of Town" (Ch. 11); "Paperclip Wasteland" (Ch. 12);
"That's Professor *Dog to You" (Ch. 14)*

You may notice, with a glare, that I've devoted four chapters to optimization, and a scant few paragraphs (in Ch. 14) to related rates. I apologize to all the related rates aficionados out there; I just didn't find good stories to house them. As for the emphasis on optimization, which is the motivating application of differential calculus, I offer not one iota of apology.

Rolle's Theorem and the Mean Value Theorem: *"The Curve's Last Laugh"*
(Ch. 13); "A Towering Baklava of Abstractions" (Ch. 26)

Although it name-drops Rolle's theorem, Ch. 13 is mostly another case study in optimization. Ch. 26 presents the MVT for derivatives and integrals side by side; toward the end, it also discusses (indeed, mocks) the intermediate value theorem. I'm sure this hodge-podge will frustrate good-hearted teachers seeking to follow the traditional sequence, but that's kind of my point: the traditional sequence is just one possible approach to this material, an approach that privileges a somewhat ahistorical notion of "rigor."

I'm honestly not sure what I'll do next time I teach this material. But I plan to emphasize that the MVT, IVT, and their ilk became central only after Fourier's work raised deep questions about convergence, thereby creating an intellectual need that delta-epsilonic rigor would satisfy.

Differential Equations: *"The Unauthorized Biography of a Fad" (Ch. 7)*

This chapter touches on several topics that deserve individual attention: (1) exponential growth; (2) inflection points; and (3) differential equations. Alas, in calendrical practice, these topics often land months or even semesters apart, which makes this chapter a rather inconvenient stew. Ah well—it goes to show that mathematics "in the wild" does not observe our national borders.

Definition of an Integral: *"In Literary Circles" (Ch. 16);*
"War and Peace and Integrals" (Ch. 17); "If Pains Must Come" (Ch. 23)

Compared to the derivative, I find the integral subtler and more elusive. The phrase "area under the curve" feels more abridged to me, and less honest, than "instantaneous rate of change."

Thus, after Ch. 16 lays a little groundwork, Ch. 17 and Ch. 23 explore a kind of fuzzy metaphorical integral. This is not much good for completing your homework, but it can perhaps serve as a conceptual touchstone. I find that geometric properties of the integral (e.g., that $\int_a^b f(x)\,dx + \int_b^c g(x)\,dx = \int_a^c f(x)\,dx$) emerge easily enough in such settings.

Riemann Sums: *"Riemann City Skyline" (Ch. 18)*

As a teacher, I suspect the best approach to Riemann sums is to compute one or two, very carefully, and then retire them from use. The sigma machinery is cumbersome to operate, especially if your algebra is the slightest bit wobbly. The inconvenience of the approach motivates the search for a shortcut, which will arrive in the glorious form of the fundamental theorem.

As a writer, however, I decided to indulge my fondness for analysis (or what little I know of it). The Dirichlet function is a favorite; it is the simplest example I know for revealing the shortcomings of the Riemann sum and the necessity of the Lebesgue integral. (Admittedly, it rests on the "intuition" that the rational numbers constitute a set of measure zero, one of the most famously counterintuitive results in elementary analysis.)

Fundamental Theorem: *"A Great Work of Synthesis" (Ch. 19)*

When I first taught calculus, we spent a week computing definite integrals via geometric methods, then a week computing indefinite integrals (i.e., antiderivatives). Both use integral notation, but as far as the students knew, these were two distinct uses, wholly unrelated. After this dull circus, I shouted, "Abracadabra! Actually, they *are* related!"

I had turned the fundamental theorem of calculus into the world's worst birthday surprise.

These days, I don't delay the fundamental theorem a moment longer than I need to. It's like in *When Harry Met Sally*: "When you realize you want to spend the rest of your life with somebody, you want the rest of your life to start as soon as possible."

Numerical Integration: *"1994, the Year Calculus Was Born" (Ch. 22);* *"Scenes from an Impossibility" (Ch. 28)*

I am not an engineer, or a diabetes researcher, or anything so practical, but it's my understanding that numerical integration is wildly useful across the sciences, and probably deserves more emphasis than the typical calculus course gives it. This is especially true now that algebraic software is getting so good at computing antiderivatives, thereby relieving us from the need to master 1001 integration techniques.

Integration Techniques: *"What Happens under the Integral Sign Stays under the Integral Sign" (Ch. 20)*

In this chapter, I attempt to give the taste and texture of integration without actually computing any integrals. This is a silly goal, and perhaps unattainable, but intrinsic to the purpose of the book, so here we are. It's not that I don't value computation, by the way; a central purpose of anything called "calculus" is to make calculation easier, more fluid, more brainless. I'm just not a good enough storyteller to fashion a compelling narrative out of trigonometric substitution.

Constants of Integration: *"Discarding Existence with a Flick of His Pen"* (Ch. 21)

Again, you can see my fondness for kinematics. I like introducing constants of integration by integrating a velocity function, where the +C has a clear physical meaning as the position at $t = 0$.

Solids of Revolution: *"Fighting with the Gods" (Ch. 24);* *"From Spheres Unseen" (Ch. 25); "Gabriel, Blow Your Trumpet" (Ch. 27)*

I think solids of revolution make a lovely conclusion to a first calculus course. They're visually striking, geometrically rich, stupefyingly technical, and they give you occasion to ramble about Archimedes and the archangel Gabriel (who has been portrayed on film by both Christopher Walken and Tilda Swinton, the two strangest actors in cinema. I know that fact doesn't really belong here, but I had nowhere else to put it, and couldn't bear omitting it from the book).

BIBLIOGRAPHY

MOMENTS

I. The Fugitive Substance of Time

- Aristotle, *Physics*. Translated by R. P. Hardie and R. K. Gaye. The Internet Classics Archives by Daniel C. Stevenson, Web Atomics, 1994–2000. http://classics.mit.edu/Aristotle/physics.mb.txt.
- Borges, Jorge Luis. "The Secret Miracle." *Collected Fictions*. Translated by Andrew Hurley. New York: Penguin Books, 1999.
- Evers, Liz. *It's About Time: From Calendars and Clocks to Moon Cycles and Light Years—A History*. London: Michael O'Mara Books, 2013.
- Gleick, James. *Time Travel: A History*. New York: Vintage Books, 2017.
- Joseph, George Gheverghese. *The Crest of the Peacock: Non-European Roots of Mathematics*. 3rd ed. Princeton, NJ: Princeton University Press, 2010.
- Mazur, Barry. "On Time (In Mathematics and Literature)." 2009. http://www.math.harvard.edu/~mazur/preprints/time.pdf.
- Stock, St. George William Joseph. *Guide to Stoicism*. Tredition Classics, 2012.
- Wolfe, Thomas. *Of Time and the River: A Legend of Man's Hunger in His Youth*. New York: Scribner Classics, 1999.

II. The Ever-Falling Moon

My tremendous thanks to Viktor Blåsjö, whose work (e.g., *History of Mathematics* and *Intuitive Infinitesimal Calculus* via IntellectualMathematics.com) helped to shape and inspire this chapter. As he points out, the way I present Newton's argument—first assuming that the inverse square law holds, and then deducing the moon's orbital period—is a sort of inside-out version of Newton's original.

"The orbital period is known, of course," explains Blåsjö, "and it is the distance the moon would fall in one second that is the mystery, the thing we need to figure out through indirect reasoning since there is no way of measuring it by experiment. This comes out in agreement with the inverse square law of gravitation (which is independently confirmed for instance by predicting elliptical planetary orbits)."

• Connor, Steve. "The Core of Truth behind Sir Isaac Newton's Apple." *Independent*, January 18, 2010. https://www.independent.co.uk/news/science/the-core-of-truth-behind -sir-isaac-newtons-apple-1870915.html.

• Epstein, Julia L. "Voltaire's Myth of Newton." *Pacific Coast Philology* 14 (October 1979): 27–33.

• Gleick, James. *Isaac Newton*. New York: Vintage Books, 2004.

• Gregory, Frederick. "Newton, the Apple, and Gravity." Department of History, University of Florida, 1998. http://users.clas.ufl.edu/fgregory/Newton_apple.htm.

• ———. "The Moon as Falling Body." Department of History, University of Florida, 1998. http://users.clas.ufl.edu/fgregory/Newton_moon2.htm.

• Keesing, Richard. "A Brief History of Isaac Newton's Apple Tree." University of York, Department of Physics. https://www.york.ac.uk/physics/about/newtonsappletree/.

• Moore, Alan. "Alan Moore on William Blake's Contempt for Newton." Royal Academy, December 5, 2014. https://www.royalacademy.org.uk/article/william-blake-isaac -newton-ashmolean-oxford.

• Voltaire. *Letters on England*. Translated by Henry Morley. Transcribed from the 1893 Cassell & Co. edition. https://www.gutenberg.org/files/2445/2445-h/2445-h.htm.

Truths are illusions which we
have forgotten are illusions –

they are metaphors that have become worn
out and have been drained of sensuous force,

coins which have lost their embossing and are now
considered as metal and no longer as coins.

—FRIEDRICH NIETZSCHE

III. The Fleeting Joys of Buttered Toast

- Berkeley, George. *The Analyst*, edited by David R. Wilkins, 2002. Based on the original 1734 edition. https://www.maths.tcd.ie/pub/HistMath/People/Berkeley/Analyst/Analyst.pdf.
- Frost, Robert. "Education by Poetry." *Amherst Graduates' Quarterly* (February 1931). http://www.en.utexas.edu/amlit/amlitprivate/scans/edbypo.html.

IV. The Universal Language

- Atiyah, Michael. "The Discrete and the Continuous from James Clerk Maxwell to Alan Turing." Lecture presented at the 5th Annual Heidelberg Laureate Forum, September 29, 2017.
- Bardi, Jason Socrates. *The Calculus Wars: Newton, Leibniz, and the Greatest Mathematical Clash of All Time*. New York: Basic Books, 2007.
- Mazur, Barry. "The Language of Explanation." Essay written for the University of Utah Symposium in Science and Literature, November 2009. http://www.math.harvard .edu/~mazur/papers/Utah.3.pdf.
- Wolfram, Stephen. "Dropping In on Gottfried Leibniz." In *Idea Makers: Personal Perspectives on the Lives and Ideas of Some Notable People*. Champaign, IL: Wolfram Media, 2016. http://blog.stephenwolfram.com/2013/05/dropping-in-on-gottfried-leibniz/.

V. When the Mississippi River Ran a Million Miles Long

My thanks to Professor Tatem for his kind response to my email inquiry, clarifying that the extrapolation was, indeed, tongue-in-cheek.

- Ellenberg, Jordan. *How Not to Be Wrong: The Power of Mathematical Thinking.* New York: Penguin Books, 2014.
- Tatem, Andrew J., Carlos A. Guerra, Peter M. Atkinson, and Simon I. Hay. "Momentous Sprint at the 2156 Olympics?" *Nature* 431, no. 525 (September 30, 2004).
- Twain, Mark. *Life on the Mississippi.* Boston: James R. Osgood, 1883. https://www.gutenberg.org/files/245/245-h/245-h.htm.

VI. Sherlock Holmes and the Bicycle of Misdirection

Huge thanks to Dan Anderson for the Desmos app I used to generate the bike tracks!

- Bender, Edward A. "Sherlock Holmes and the Bicycle Tracks." University of California, San Diego. http://www.math.ucsd.edu/~ebender/87/bicycle.pdf.
- Doyle, Arthur Conan. "The Adventure of the Priory School." In *The Return of Sherlock Holmes.* New York: McClure, Phillips & Co., 1905. https://en.wikisource.org/wiki/The_Adventure_of_the_Priory_School.
- Duchin, Moon. "The Sexual Politics of Genius." University of Chicago, 2004. https://mduchin.math.tufts.edu/genius.pdf.
- O'Connor, J. J., and E. F. Robertson. "James Moriarty." School of Mathematics and

Statistics, University of St. Andrews. http://www-groups.dcs.st-and.ac.uk/history /Biographies/Moriarty.html.

• Roberts, Siobhan. *Genius at Play: The Curious Life of John Horton Conway*. New York: Bloomsbury, 2015.

VII. The Unauthorized Biography of a Fad

A shout-out to Rebecca Jackman, my high school chemistry teacher, who is blameless for any errors in my discussion of autocatalysis and responsible for any non-errors.

• Jones, Jamie. "Models of Human Population Growth." Monkey's Uncle: Notes on Human Ecology, Population, and Infectious Disease, April 7, 2011. http://monkeysuncle .stanford.edu/?p=933. Jones provides the "mechanistic vs. phenomenological" framework.

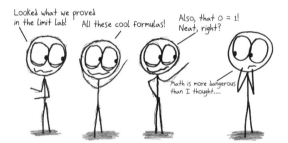

VIII. What the Wind Leaves Behind

• Brown, Kevin. "The Limit Paradox." Math Pages. https://www.mathpages.com/home /kmath063.htm. I found Professor Brown's discussion clarifying and essential. I also like that his name appears nowhere on his site, giving it a disembodied "the voice of mathematics" atmosphere.

• Dunham, William. *The Calculus Gallery: Masterpieces from Newton to Lebesgue*. Princeton, NJ: Princeton University Press, 2008. I grazed upon Dunham's book in January 2016. His insights about the history of analysis circulated through my mental stomachs for several years. This book is the milk, so to speak.

I walk a path no mind can heed
I run a race yet go no speed
and even as you hear this verse
a sudden corner I traverse.

— What am I?

The plot of "Lost"?

IX. Do the Dusty Dance

- Blåsjö, Viktor. "Attitudes toward Intuition in Calculus Textbooks." Paper forthcoming, 2019. In this paper, Blåsjö pushes back against the standard narrative that the Weierstrass function was the "death of intuition." Worth a read if you're into this history.
- Dunham, William. *The Calculus Gallery.*
- Fowler, Michael. "Brownian Motion." University of Virginia, 2002. http://galileo.phys .virginia.edu/classes/152.mf1i.spring02/BrownianMotion.htm.
- Isaacson, Walter. *Einstein: His Life and Universe.* New York: Simon & Schuster, 2007.
- Poincaré, Henri. "L'Oeuvre Mathématique de Weierstrass." *Acta Mathematica* 22 (1899): 1–18. https://projecteuclid.org/download/pdf_1/euclid.acta/1485882041. I don't speak French, but happily, Google Translate does.
- Yeo, Dominic. "Remarkable Fact about Brownian Motion #1: It Exists." *Eventually Almost Everywhere.* January 22, 2012. https://eventuallyalmosteverywhere.wordpress .com/2012/01/22/remarkable-fact-about-brownian-motion-1-it-exists/.

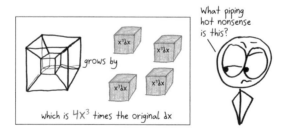

X. The Green-Haired Girl and the Superdimensional Whorl

- Roberts, Siobhan. *King of Infinite Space: Donald Coxeter, the Man Who Saved Geometry.* New York: Walker, 2006. Pilfered for quotes and insights on the history of geometric thinking (and its devastation at the hands of that gorgeous monster Bourbaki).
- St. Clair, Margaret. "Presenting the Author." *Fantastic Adventures,* November 1946: 2–5.
- ———. "Aleph Sub One," *Startling Stories,* January 1948: 62–69. I confess, the story deals only with expanding $(a+b)^n$ for $n = 2, 3$, and 4; the idea of applying these expansions to the derivative formulas is my own cheeky extrapolation.

- Thompson, Silvanus P. *Calculus Made Easy: Being a Very-Simplest Introduction to Those Beautiful Methods of Reckoning Which Are Generally Called By the Terrifying Names of the Differential Calculus and the Integral Calculus.* 2nd ed. London: Macmillan, 1914. The book, available free online, is even more delightful than its title. See particularly the second chapter, "On Different Degrees of Smallness." https://www.gutenberg.org/files/33283/33283-pdf.pdf.

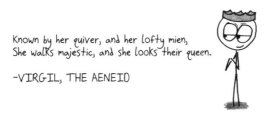

Known by her quiver, and her lofty mien,
She walks majestic, and she looks their queen.

−VIRGIL, THE AENEID

XI. The Princess on the Edge of Town

- Lendering, Jona. "Carthage." *Livius.org: Articles on Ancient History.* http://www.livius.org/articles/place/carthage/.
- ———. "The Founding of Carthage." *Livius.org: Articles on Ancient History.* http://www.livius.org/sources/content/the-founding-of-carthage/.
- Virgil. *The Aeneid.* Translated by John Dryden. http://classics.mit.edu/Virgil/aeneid.html.

I thought you were watching some Netflix to relax...?

MY SEARCH FOR THE MOST RELAXING FILM IS DRIVING ME TO AN EARLY GRAVE!

XII. Paperclip Wasteland

- Bostrom, Nick. "Ethical Issues in Advanced Artificial Intelligence." https://nickbostrom.com/ethics/ai.html.
- Chiang, Ted. "Silicon Valley Is Turning into Its Own Worst Fear." *BuzzFeed News,* December 18, 2017. https://www.buzzfeednews.com/article/tedchiang/the-real-danger-to-civilization-isnt-ai-its-runaway.
- Fry, Hannah. *Hello World: Being Human in the Age of Algorithms.* New York: W. W. Norton, 2018.
- Whitman, Walt. "Song of Myself." 1855. *Leaves of Grass* (final "Death-Bed" edition, 1891–92) (David McKay, 1892).

- Yudkowsky, Eliezer. "There's No Fire Alarm for Artificial General Intelligence." Machine Intelligence Research Institute, October 13, 2017. https://intelligence .org/2017/10/13/fire-alarm/.
- Yudkowsky, Eliezer. "Artificial Intelligence as a Positive and Negative Factor in Global Risk." In *Global Catastrophic Risks*, edited by Nick Bostrom and Milan M. Ćirković, 308–345. New York: Oxford University Press, 2008. http://intelligence.org/files /AIPosNegFactor.pdf.
- Zunger, Yonatan. "The Parable of the Paperclip Maximizer." Hacker Noon, July 24, 2017. https://hackernoon.com/the-parable-of-the-paperclip-maximizer-3ed4cccc669a.

1978: The Laffer Curve Begins to Question the Burden Jude Wanniski Has Placed On It

XIII. *The Curve's Last Laugh*

- Appelbaum, Binyamin. "This Is Not Arthur Laffer's Famous Napkin." *New York Times*, October 13, 2017. https://www.nytimes.com/2017/10/13/us/politics/arthur-laffer-napkin-tax-curve.html.
- Bernstein, Adam. "Jude Wanniski Dies; Influential Supply-Sider." *Washington Post*, August 31, 2005. http://www.washingtonpost.com/wp-dyn/content/article/2005/08/30 /AR2005083001880.html.
- Chait, Jonathan. "Prophet Motive." *New Republic*, March 30, 1997. https://newrepublic.com/article/93919/prophet-motive.
- ———. "Flight of the Wingnuts: How a Cult Hijacked American Politics." *New Republic*, September 10, 2007. http://www.wright.edu/~tdung/How_supply_eco_hijacked_US_Politics.pdf.
- Gardner, Martin. "The Laffer Curve." In *Knotted Doughnuts and Other Mathematical Entertainments*, 257–71. New York: W. H. Freeman, 1986.
- Laffer, Arthur. "The Laffer Curve: Past, Present, and Future." Heritage Foundation, June 1, 2004. https://www.heritage.org/taxes/report/the-laffer-curve-past-present-and-future.
- "Laffer Curve." Chicago Booth: IGM Forum. June 26, 2012. http://www.igmchicago.org/ surveys/laffer-curve. Quotes drawn from Austan Goolsbee, Bengt Holmström, Kenneth Judd, Anil Kashyap, and Richard Thaler.
- "Laffer Curve Napkin." National Museum of American History. http://americanhistory .si.edu/collections/search/object/nmah_1439217.
- Miller, Stephen. "Jude Wanniski, 69, Provocative Crusader for Supply-Side Economics."

New York Sun, August 31, 2005. https://www.nysun.com/obituaries/jude-wanniski-69-provocative-crusader-for-supply/19386/.

• Moore, Stephen. "The Laffer Curve Turns 40: The Legacy of a Controversial Idea." *Washington Post*, December 26, 2014. https://www.washingtonpost.com/opinions/the-laffer-curve-at-40-still-looks-good/2014/12/26/4cded164-853d-11e4-a702-fa31ff4ae98e_story.html.

• Oliver, Myrna. "Jude Wanniski, 69; Journalist and Political Consultant Pushed Supply-Side Economics." *Los Angeles Times*, August 31, 2005. http://articles.latimes.com/2005/aug/31/local/me-wanniski31.

• "The 100 Best Non-Fiction Books of the Century." *National Review*. May 3, 1999. https://www.nationalreview.com/1999/05/non-fiction-100/.

• Shields, Mike. "The Brain behind the Brownback Tax Cuts." Kansas Health Institute News Service, August 14, 2012. https://www.khi.org/news/article/brain-behind-brownback-tax-cuts.

• Starr, Roger. "The Way the World Works, by Jude Wanniski." *Commentary*, September 1978. https://www.commentarymagazine.com/articles/the-way-the-world-works-by-jude-wanniski/.

• Wanniski, Jude. "The Mundell-Laffer Hypothesis—a New View of the World Economy." *Public Interest* 39 (1975) 31–52. https://www.nationalaffairs.com/storage/app/uploads/public/58e/1a4/be4/58e1a4be4e900066158619.pdf.

• ———. "Taxes, Revenues, and the 'Laffer Curve.'" *Public Interest* 50 (1978): 3–16. https://www.nationalaffairs.com/storage/app/uploads/public/58e/1a4/c54/58e1a4c549207669125935.pdf.

XIV. That's Professor *Dog* to You

My tremendous gratitude to Professor Tim Pennings for sharing his time (and his press clippings!) with me for this chapter. It is my honor and duty to carry forward the story of Elvis.

• Bolt, Michael, and Daniel C. Isaksen. "Dogs Don't Need Calculus." *College Mathematics Journal* 41, no. 10 (January 2010): 10–16. https://www.maa.org/sites/default/files/Bolt2010.pdf.

• "CNN Student News Transcript: September 26, 2008." http://www.cnn.com/2008/LIVING/studentnews/09/25/transcript.fri/index.html.

• Dickey, Leonid. "Do Dogs Know Calculus of Variations?" *College Mathematics Journal* 37, no. 1 (January 2006): 20–23. https://www.maa.org/sites/default/files/Dickey-CMJ-2006.pdf.
• "Do Dogs Know Calculus? The Corgi Might." National Purebred Dog Day, March 15, 2016. https://nationalpurebreddogday.com/dogs-know-calculus-corgi-knows/.
• Minton, Roland, and Timothy J. Pennings. "Do Dogs Know Bifurcations?" *College Mathematics Journal* 38, no. 5 (November 2007): 356–61. https://www.maa.org/sites/default/files/pdf/upload_library/22/Polya/minton356.pdf.
• Pennings, Timothy J. "Do Dogs Know Calculus?" *College Mathematics Journal* 34, no. 3 (May 2003): 178–82. https://www.jstor.org/stable/3595798.
• Perruchet, Pierre, and Jorge Gallego, "Do Dogs Know Related Rates Rather Than Optimization?" *College Mathematics Journal* 37, no. 1 (January 2006): 16–18. https://www.maa.org/sites/default/files/pdf/mathdl/CMJ/cmj37-1-016-018.pdf.
• Thurber, James. *Thurber's Dogs: A Collection of the Master's Dogs, Written and Drawn, Real and Imaginary, Living and Long Ago.* New York: Simon & Schuster, 1955.

Gottfried
Leibniz

LOCATION:
Hannover

OBJECTIVE:
Get out of
Hannover

C.V.
• I led a legal reform project to unify patchwork local laws into a single coherent system.
• I proposed modernizing institutions such as a socioeconomic census, centralized state archives, and subsidies for optimal farming practices.
• I served as chief mediator in an effort to reconcile feuding religious groups. (Don't ask me how it went.)
• I advocated for government-provided healthcare, including a proactive approach to preventing epidemics.
• I helped to found a scientific academy aiming "to improve not only the arts and sciences, but also agriculture, manufacture, commerce, and, in a word, whatever is useful in the support of life."
• I developed an influential theory of existence.

XV. Calculemus!

• Arnol'd, Vladimir. *Huygens and Barrow, Newton and Hooke.* Translated by Eric J. F. Primrose. Basel: Birkhäuser Verlag, 1990.
• Bardi, Jason Socrates. *The Calculus Wars.*
• Goethe, Norma B.; Philip Beeley, and David Rabouin, eds. *G. W. Leibniz, Interrelations between Mathematics and Philosophy.* New York: Springer, 2015.
• Grossman, Jane, Michael Grossman, and Robert Katz. *The First Systems of Weighted Differential and Integral Calculus.* Rockport, MA: Archimedes Foundation, 1980. The Gauss quote comes from page ii.
• Kafka, Franz. *The Trial.* London: Vintage, 2005. Translated by Willa and Edwin Muir.
• Wolfram, Stephen. "Dropping In on Gottfried Leibniz."

ETERNITIES

XVI. In Literary Circles

• Borges, Jorge Luis. "Pascal's Sphere." In *Other Inquisitions, 1937–1952*. Translated by Ruth L. C. Simms. Austin: University of Texas Press, 1975.

• Dauben, Joseph W. "Chinese Mathematics." In *The Mathematics of Egypt, Mesopotamia, China, India, and Islam: A Sourcebook*, edited by Victor Katz, 186–384. Princeton, NJ: Princeton University Press, 2007.

• Donne, John. "A Valediction Forbidding Mourning." In *Songs and Sonnets*.

• Hidetoshi, Fukagawa, and Tony Rothman. *Sacred Mathematics: Japanese Temple Geometry*. Princeton, NJ: Princeton University Press, 2008.

• Joseph, George Gheverghese. *The Crest of the Peacock*.

• Ken'ichi, Sato. "Chapter 2: Seki Takakazu." In *Japanese Mathematics in the Edo Period*. National Diet Library of Japan, 2011. http://www.ndl.go.jp/math/e/s1/2.html.

• Strogatz, Steven. *The Joy of x: A Guided Tour of Math, from One to Infinity*. New York: Mariner Books, 2013.

• Szymborska, Wislawa. "Pi." In *Poems New and Collected*. New York: Mariner Books, 2000.

Writer Types, According to Isaiah Berlin

FOX

a writer whose "thought is scattered or diffused, moving on many levels" (e.g., Shakespeare, Aristotle)

HEDGEHOG

a writer committed to a "single, universal, organising principle" (e.g., Plato, Dante)

TOLSTOY

I'm a hedgehog!! Why do I have to keep explaining this???

XVII. War and Peace and Integrals

- Berlin, Isaiah. *The Hedgehog and the Fox*, edited by Henry Hardy. Princeton, NJ: Princeton University Press, 2013. Original essay published in 1951.
- Dirda, Michael. "If the World Could Write…" *Washington Post*. October 28, 2007. http://www.washingtonpost.com/wp-dyn/content/article/2007/10/25/AR2007102502856.html.
- Tolstoy, Leo. *War and Peace*. 1869.

MASCOT AUDITIONS FOR THIS BOOK

Tangent Line
too tangential

Infinitesimal
too hard to see

Derivative
too derivative (of the tangent's look)

Sphere in Cylinder
under contract with Aristotle

Solid of Revolution
too revolutionary

Leo Tolstoy
too intense, man, way too intense

Local Maximum
nowhere to go but down

Elvis the Corgi
too distractingly adorable

Riemann Sum
WINNER!!

XVIII. Riemann City Skyline

- Corrigan, Maureen. *Leave Me Alone, I'm Reading: Finding and Losing Myself in Books.* New York: Random House, 2005.
- Dunham, William. *The Calculus Gallery.*
- Hamill, Pete. "A New York Writer's Take on How His City Has Changed," *National Geographic*, November 15, 2015. https://www.nationalgeographic.com/new-york-city-skyline-tallest-midtown-manhattan/article.html.
- Lindner, Christoph. "New York Vertical: Reflections on the Modern Skyline." *American Studies* 47, no. 1 (Spring 2006): 31–52. https://core.ac.uk/download/pdf/148648368.pdf.
- Rand, Ayn. *The Fountainhead.* New York: New American Library, 1994.

THE WITCH OF AGNESI

XIX. A Great Work of Synthesis

- Knill, Oliver. "Some Fundamental Theorems in Mathematics." Harvard University. http://www.math.harvard.edu/~knill/graphgeometry/papers/fundamental.pdf.
- Mazzotti, Massimo. *The World of Maria Gaetana Agnesi, Mathematician of God.* Baltimore: Johns Hopkins University Press, 2007.
- Navarro, Joaquin. "Women in Maths: From Hypatia to Emmy Noether." In *Everything Is Mathematical.* Barcelona: RBA Coleccionables, 2013.
- Ouellette, Jennifer. *The Calculus Diaries: How Math Can Help You Lose Weight, Win in Vegas, and Survive a Zombie Apocalypse.* New York: Penguin Books, 2010.

XX. What Happens under the Integral Sign Stays under the Integral Sign

My thanks to Inna Zakharevich for a helpful and enjoyable email exchange.

• Feynman, Richard P. *"Surely You're Joking, Mr. Feynman!": Adventures of a Curious Character.* New York: W. W. Norton, 1985.
• Gaither, Carl C. and Alma E. Cavazos-Gaither, eds. Gaither's Dictionary of Scientific Quotations. New York: Springer Science & Business Media, 2008.
• Gleick, James. *Genius: The Life and Science of Richard Feynman.* New York: Pantheon Books, 1992.
• Ouellette, Jennifer. *The Calculus Diaries.*
• Ury, Logan R. "Burden of Proof." *Harvard Crimson.* December 6, 2006. https://www.thecrimson.com/article/2006/12/6/burden-of-proof-at-1002-am/.
• Zakharevich, Inna. "Another Derivation of Euler's Integral Formula." Reported by Noam D. Elkies. Harvard University. http://www.math.harvard.edu/~elkies/Misc/innaz.pdf.

XXI. Discarding Existence with a Flick of His Pen

I owe a tremendous thank-you to Paul Ramond, a PhD student in physics who coached me on cosmology via Skype. Any remaining errors are wholly my own.

• Einstein, Albert. "Cosmological Considerations in the General Theory of Relativity." Translated by W. Perrett and G. B. Jeffery. Reprinted from *The Principle of Relativity,* 175–89. New York: Dover, 1952. https://einsteinpapers.press.princeton.edu/vol6-trans/433.
• Harvey, Alex, "The Cosmological Constant." New York University, November 23, 2012. https://arxiv.org/pdf/1211.6337.pdf.

• Isaacson, Walter. *Einstein.*

• Janzen, Daryl. "Einstein's Cosmological Considerations." University of Saskatchewan. February 13, 2014. https://arxiv.org/pdf/1402.3212.pdf.

• Munroe, Randall. "The Space Doctor's Big Idea." *New Yorker*, November 18, 2015.

• Ohanian, Hans. *Einstein's Mistakes: The Human Failings of Genius.* New York: W. W. Norton & Company, 2008.

• O'Raifeartaigh, C., and B. McCann. "Einstein's Cosmic Model of 1931 Revisited: An Analysis and Translation of a Forgotten Model of the Universe." Waterford Institute of Technology. https://arxiv.org/ftp/arxiv/papers/1312/1312.2192.pdf.

• O'Raifeartaigh, Cormac, Michael O'Keeffe, Werner Nahm, and Simon Mitton. "Einstein's 1917 Static Model of the Universe: A Centennial Review." https://arxiv.org/ftp/arxiv/papers/1701/1701.07261.pdf.

• Rovelli, Carlo. *Seven Brief Lessons on Physics.* New York: Riverhead Books, 2016.

• Straumann, Norbert. "The History of the Cosmological Constant Problem." Institute for Theoretical Physics, University of Zurich, August 13, 2001. https://arxiv.org/pdf/gr-qc/0208027.pdf.

XXII. 1994, the Year Calculus Was Born

• Łaba, Izabella. "The Mathematics of Wheel Reinvention." *The Accidental Mathematician.* January 18, 2016. https://ilaba.wordpress.com/2016/01/18/the-mathematics-of-wheel-reinvention/.

• "Letters." *Diabetes Care* 17, no. 10 (October 1994): 1223–27. Authors of quoted letters include Ralf Bender; Thomas Wolever; Jane Monaco and Randy Anderson; and Mary Tai.

• "Medical Researcher Discovers Integration, Gets 75 Citations." *An American Physics Student in England.* March 19, 2007. https://fliptomato.wordpress.com/2007/03/19/medical-researcher-discovers-integration-gets-75-citations/.

• Ossendrijver, Mathieu. "Ancient Babylonian Astronomers Calculated Jupiter's Position from the Area under a Time-Velocity Graph." *Science* 351, no. 6272 (January 29, 2016): 482–84.

• Tai, Mary. "A Mathematical Model for the Determination of Total Area under Glucose Tolerance and Other Metabolic Curves." *Diabetes Care* 17, no. 2 (February 1994): 152–54.

• Trefethen, Lloyd N. "Numerical Analysis." In *Princeton Companion to Mathematics*, edited by Timothy Gowers, June Barrow-Green, and Imre Leader. Princeton, NJ: Princeton University Press, 2008. http://people.maths.ox.ac.uk/trefethen/NAessay.pdf.

• Wolever, Thomas. "How Important Is Prediction of Glycemic Responses?" *Diabetes Care* 12, no. 8 (September 1989): 591–93.

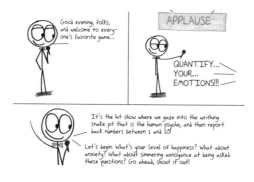

XXIII. If Pains Must Come

• Bentham, Jeremy. *An Introduction to the Principles of Morals and Legislation.* Adapted by Jonathan Bennett. https://www.earlymoderntexts.com/assets/pdfs/bentham1780.pdf.
• Bradbury, Ray. *Bradbury Speaks: Too Soon From the Cave, Too Far from the Stars.* New York: William Morrow, 2006.
• Dickinson, Emily. "Bound—a Trouble." (No. 269.) https://en.wikisource.org/wiki /Bound_—_a_trouble_—.
• Frost, Robert. "Happiness Makes Up in Height for What It Lacks in Length." In *The Poetry of Robert Frost: The Collected Poems, Complete and Unabridged.* New York: Henry Holt and Co., 1999.
• Jevons, William Stanley. "Brief Account of a General Mathematical Theory of Political Economy." *Journal of the Royal Statistical Society, London* XXIX (June 1866): 282–87. https://www.marxists.org/reference/subject/economics/jevons/mathem.htm.
• Kahneman, Daniel, Barbara L. Fredrickson, Charles A. Schreiber, and Donald A. Redelmeier. "When More Pain Is Preferred to Less: Adding a Better End." *Psychological Science* 4, no. 6 (November 1993): 401–5.
• Mill, John Stuart. *Utilitarianism* (edited by George Sher). Indianapolis: Hackett Publishing Co., 2002. Page 10.
• Singer, Peter. *Animal Liberation: Updated Edition.* New York: Harper Perennial, 2009.

XXIV. Fighting with the Gods

- Brown, Kevin. "Archimedes on Spheres and Cylinders." Math Pages. https://www.mathpages.com/home/kmath343/kmath343.htm.
- Leibniz, Gottfried Wilhelm Freiherr, and Antoine Arnauld. *The Leibniz-Arnauld Correspondence.* New Haven, CT: Yale University Press, 2016.
- Lockhart, Paul. *Measurement.* Cambridge, MA: Belknap Press, 2012.
- Plutarch. *Lives of the Nobel Greeks and Romans.* http://www.fulltextarchive.com /page/Plutarch-s-Lives10/#p35.
- Polster, Burkard. *Q.E.D.: Beauty in Mathematical Proof.* New York: Bloomsbury, 2004.
- Polybius. *Universal History, Book VIII.* Excerpted from *The Rise of the Roman Empire,* translated by Ian Scott-Kilvert. New York: Penguin Books, 1979. https://www.math.nyu.edu/~crorres/Archimedes/Siege/Polybius.html.
- Rorres, Chris. "Death of Archimedes: Sources." New York University. https://www.math.nyu.edu/~crorres/Archimedes/Death/Histories.html.
- Sharratt, Michael. *Galileo: Decisive Innovator.* Cambridge, UK: Cambridge University Press, 1994. Page 52.
- Whitehead, Alfred North. *An Introduction to Mathematics.* New York: Henry Holt and Company, 1911.

FUN FACT: Instead of stacks of disks, you can view solids of revolution as onions, where each layer is a rolled-up paper cylinder.

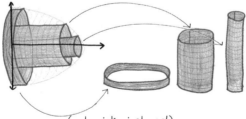

(not a joke, just cool)

XXV. From Spheres Unseen

Thanks to my Twitter pals Ben Blum-Smith (@benblumsmith) and Mike Lawler (@mike-andallie) for helping me out with the volume of the 4D sphere.

- Abbott, Edwin. *Flatland: A Romance of Many Dimensions.* 1884.
- Strogatz, Steven. *The Calculus of Friendship: What a Teacher and a Student Learned about Life while Corresponding about Math.* Princeton, NJ: Princeton University Press, 2009.

David Foster Wallace Says...

On Leibniz: "A lawyer/diplomat/ courtier/philosopher for whom math was sort of an offshoot hobby." A footnote adds: "Surely we all hate people like this."

"Aristotle manages to be sort of grandly and breathtakingly wrong, always and everywhere, when it comes to ∞."

On the real number line: "99.999...% empty space, rather like DQ ice cream or the universe itself."

On using calculus maneuvers to "solve" Zeno's paradox: "Complex, formally sexy, technically correct, and deeply trivial."

On the Newton vs. Leibniz priority dispute: "The idea of exclusive or even dual credit is absurd, as is the notion that what's now called the calculus comprises any one invention."

On doing math whose logical foundations are unclear: "A stock-market bubble." He also says such math is "trying to tie its shoes on the run."

On Georg Cantor: "A completely average-looking bourgeois German from the era of starched collars and fire-hazard beards."

On Karl Weierstrass: "Conspicuous among mathematicians for being physically large, a gifted athlete, an inveterate partier and blowoff in college, indifferent to music (most mathematicians are fiends for music), and a cheery, non-neurotic, gregarious, wholly good and much-loved fellow. He's also widely regarded as the greatest math teacher of the century, even though he never published his lectures or even let his students take notes."

XXVI. A Towering Baklava of Abstractions

- Arnold, Vladimir. "On Teaching Mathematics." Translated by A. V. Goryunov. *Russian Mathematical Surveys* 53, no. 1 (1998): 229–36.
- Cheng, Eugenia. *Beyond Infinity: An Expedition to the Outer Limits of Mathematics.* New York: Basic Books, 2017.
- Ellenberg, Jordan. *How Not to Be Wrong.*
- Kakutani, Michiko. "A Country Dying of Laughter. In 1,079 Pages." *New York Times*, February 13, 1996. https://www.nytimes.com/1996/02/13/books/books-of-the-times-a-country-dying-of-laughter-in-1079-pages.html.
- Max, Daniel T. *Every Love Story is a Ghost Story: A Life of David Foster Wallace.* New York: Viking, 2012.
- McCarthy, Kyle. "Infinite Proofs: The Effects of Mathematics on David Foster Wallace." *Los Angeles Review of Books*, November 25, 2012. https://lareviewofbooks.org/article/infinite-proofs-the-effects-of-mathematics-on-david-foster-wallace/.
- Papineau, David. "Room for One More." *New York Times*, November 16, 2003. http://www.nytimes.com/2003/11/16/books/room-for-one-more.html.
- Scott, A. O. "The Best Mind of His Generation." *New York Times*, September 20, 2008. https://www.nytimes.com/2008/09/21/weekinreview/21scott.html.

- Wallace, David Foster. "Tennis, Trigonometry, Tornadoes: A Midwestern Boyhood." *Harper's Magazine*, December 1991.
- ———. *Infinite Jest*. New York: Little, Brown, 1996.
- ———. "Rhetoric and the Math Melodrama." *Science* 290, no. 5500 (December 22, 2000): 2263–67.
- ———. *Everything and More: A Compact History of Infinity*. New York: W. W. Norton, 2003.

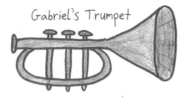

Gabriel's Trumpet

(not pictured, but Google-able: Gabriel's Wedding Cake; Gabriel's Funnel; Gabriel's Beer Glass; Gabriel's Lava Lamp; Thor's Anvil)

XXVII. Gabriel, Blow Your Trumpet

- Alexander, Amir. *Infinitesimal: How a Dangerous Mathematical Theory Shaped the Modern World*. New York: Farrar, Straus and Giroux, 2014.
- Cucić, Dragoljub. "Types of Paradox in Physics." Regional Centre for Talents Mihajlo Pupin. https://arxiv.org/ftp/arxiv/papers/0912/0912.1864.pdf.
- Gethner, Robert M. "Can You Paint a Can of Paint?" *College Mathematics Journal* 36, no. 4 (November 2005): 400–402.
- Hofstadter, Douglas. *Gödel, Escher, Bach: An Eternal Golden Braid*. New York: Basic Books, 1979.
- Smith, Wendy, and Marianne Lewis. "Leadership Skills for Managing Paradoxes." *Industrial and Organizational Psychology* 5, no. 2 (June 2012).

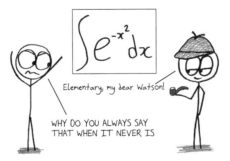

$$\int e^{-x^2} dx$$

Elementary, my dear Watson!

WHY DO YOU ALWAYS SAY THAT WHEN IT NEVER IS

VIII. Scenes from an Impossibility

My thanks to Taryn Flock for walking me through the proof of the Gaussian integral.

- Chiang, Ted. *Stories of Your Life and Others*. New York: Tom Doherty Associates, 2002.
- Oliva, Philip B. *Antioxidants and Stem Cells for Coronary Heart Disease*. Singapore: World Scientific Publishing, 2014. Page 534.

ACKNOWLEDGMENTS

The Cavalieris Whose Magic Transformed My Dubious Stack
of Infinitesimal Thoughts into a Solid and Wondrous Book:

I wish to thank the dream team of latter-day Cavalieris who helped to birth this book: the editorial wisdom of Becky Koh; the marketing and publicity expertise of Betsy Hulsebosch and Kara Thornton; the design glory of Paul Kepple, Alex Bruce, and Katie Benezra; the unflinching eyes of Melanie Gold and Elizabeth Johnson; the photographic splendor of Rayleen Tritt; the tireless genius of the whole Black Dog & Leventhal team; and the trusty guidance of Dado Derviskadic and Steve Troha, without whom this book would never have existed.

The Agnesis Whose Tutelage Led Me from Ignorance:

When the first draft of this book collapsed, it was David Klumpp who pulled me from the wreckage, dusted me off, and helped me architect a better plan. I give unbounded thanks for his help. I also owe special gratitude to the folks who gave excellent feedback at various stages: Viktor Blåsjö, Richard Bridges, Karen Carlson, John Cowan, David Litt, Doug Magowan, Jim Orlin, Jim Propp, and Katy Waldman. All remaining errors in the text are wholly and singularly my own.

The Twains and Tolstoys Whose Tales Nourished This Volume:

I'm grateful to the mathematicians who shared their stories with me, including Tim Pennings (Ch. 14) and Inna Zakharevich (Ch. 20). As for Andy Bernoff, Kay Kelm, Jonathan Rubin, and Stacey Muir, I owe them both thanks and apologies: they shared wondrous tales of a beautiful event called "the Integration Bee," but I was not a deft enough writer to make the chapter work here. Still, I pledge to share the legend of the Bee (which is Bernoff's brainchild) elsewhere. It deserves to be told.

The Gaussian Integrals Whom I Love Without Hesitation:

To my colleagues, my students, my teachers, my friends, my family, my fondest nemeses, my Twitter heroes, my Branfordians, my blog commenters, my baristas—and, most of all, to Taryn—I give my heartfelt thanks.

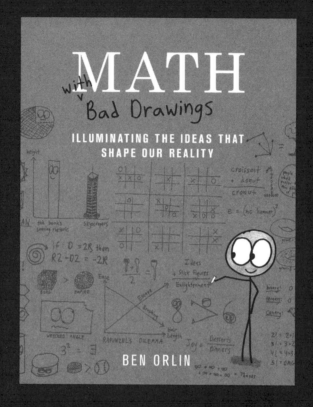